AF364247

PRODUCTION TECHNOLOGY OF DRUMSTICK
(*Moringa oleifera* Lamk.)

PRODUCTION TECHNOLOGY OF DRUMSTICK
(*Moringa oleifera* Lamk.)

Dr. R. Sri Hari Babu
Professor of Horticulture (Retd.)

Dr. (Smt.) C. Madhumathi
Principal Scientist (Hort.)

Smt. G. Jyothi
Asst. Professor (Hort)

BS Publications
A unit of **BSP Books Pvt. Ltd.**
4-4-309/316, Giriraj Lane,
Sultan Bazar, Hyderabad - 500 095

Production Technology of Drumstick (*Moringa oleifera* Lamk.)
***By* Dr. R. Sri Hari Babu, Dr. (Smt.) C. Madhumathi and Smt. G. Jyothi**

Published by:

BSP **BS Publications**
A unit of **BSP Books Pvt. Ltd.**
4-4-309/316, Giriraj Lane, Sultan Bazar,
Hyderabad - 500 095
Phone : 040 - 23445688
e-mail : info@bspbooks.net
www.bspbooks.net

ISBN: 978-93-91910-19-8 (Hardbound)

Preface

The moringa/drumstick (*Moringa oleifera* Lam.) is one of the most incredible plants gifted to mankind by nature. Its nutritional and medicinal properties have a tremendous potential to end malnutrition, starvation as well as to prevent and heal many diseases and disorders of humans as well as animals worldwide. According to Ayurveda, the ancient system of medicine of India, moringa cures 300 diseases and disorders of human beings. Moringa is truly "A miracle plant" and divine gift for nourishing and healing of humans. The outstanding feature of moringa is that all parts of the plant are useful in one way or other to mankind. This plant species possesses so many uses and special features which amazes the modern scientists.

Though it has long history it did not receive much attention for a long time. But now it is re-discovered during the second half of 20th century not only as a food plant but mainly as a medicinal plant. In the recent times the emerging economic profitability and desirable health benefits of moringa have caught the attention of the world. During the past five decades a great volume of information has been generated and accumulated in research in different parts of the world notably in India. As a token of recognition of moringa's multifarious qualities and uses, International Society of Horticulture Science has recently conducted First International Symposium on Moringa in 2015 in Philippines which was published as Acta Horticulturae No. 1158 in 2017. A few monographs on moringa from different countries have also been published. After two decades of publication of Lowell Fuglie's first publication of the book "The Miracle Tree" – *Moringa oleifera*, there has been much progress in moringa's Research and Development in response to increasing demand for new information on various topics about moringa which resulted in the publication of 2nd edition of "The Miracle Tree " *Moringa oleifera* Lam by M.C.Palada and his colleagues in 2019. A considerable effort has been made in bringing the accumulated information on package of practices on production of moringa from various sources for the benefit of all concerned with moringa, authors have focussed mainly on production technology of drumsticks (pods.)

The book presents much information on cultivation of drumstick from various sources including field research studies conducted by various research organisations worldwide, articles written by individuals with experience and knowledge about moringa as well as other books and publications cited in the book. However, the authors and publishers cannot / do not assume responsibility for the validity of all information in the book or consequences of their adoption.

The authors wish cordially to acknowledge the support and help rendered by colleagues and friends in supplying information, materials, photos in improving the information of the book and all those who worked on drumstick.

-Authors

List of Tables

List of Figures/Plates

Contents

Abbreviations

S.No.

1.	AC	Acre
2.	AM	ArbuscularMycorrhiza
3.	CAS	Controlled Atmospheric Storage
4.	CFB	Corrugated Fibre Board
5.	CM	Centimetres
6.	DAA	Days after Anthesis
7.	DAD	Di Ammonium Phosphate
8.	ESP	Exchangeable Sodium Phosphate
9.	EC	Electrical Conductivity
10.	g	Gram
11.	ha	Hectare
12.	HDP	High Density Planting
13.	HDPE	High Density Polyethylene
14.	IIHR	Indian Institute of Horticultural Research
15.	IIVR	Indian institute of Vegetable Research
16.	INM	Integrated Nutrient Management
17.	IPM	Integrated Pest Management
18.	IWM	Integrated Weed Management
19.	kg	Kilogram
20.	kvk	KrishiVignanKendram
21.	l	Litre
22.	LDPE	Low Density Polyethylene
23.	M	Metre
24.	ml	Millilitre
25.	NGO	Non-government Organization.
26.	NPK	Nitrogen, phosphorus, potassium.
27.	NSKE	Neem Seed Kernal Extract

28.	ONGC	Oil and Natural Gas Commission.
29.	PKM	Periakulum
30.	POP	Package of Practices
31.	ppm	Parts per million.
32.	%	Percentage
33.	PSB	Phosphorus Solubilizing Bacteria.
34.	PVC	Poly Vinyl Chloride
35.	Rs	Rupees
36.	SIF	Save Indian Farmers
37.	TNAU	Tamil Nadu Agricultural University
38.	USA	United States of America
39.	USD	United States Dollars
40.	VAM	Vesicular Arbuscular Mycorrhia
41.	ZECC	Zero Energy Cool Chamber

About the Authors

Dr.R. Sri Hari Babu

Born in 1945 in Andhra Pradesh, obtained B.Sc.(Ag.) degree with 1st class in 1968 from Andhra Pradesh Agricultural University, Hyderabad (Now Acharya N.G. Ranga Agricultural University, Guntur, Andhra Pradesh), M.Sc.(Ag.) Horticulture with distinction in 1970 from U.P. Agricultural University (Now G.B. Pant University of Agriculture and Technology), Pantnagar, Uttarkhand and Ph.D.(Hort) from Banaras Hindu University, Varanasi, Uttar Pradesh in 1981.

He joined A.P. Agricultural University as Horticulture Assistant at College of Agriculture, Bapatla, Andhra Pradesh in 1972. Promoted as Associate Professor in 1982 and as Senior Scientist (Professor) in 1989, worked at Bapatla, Kodur, Hyderabad, Sangareddy and Tirupati at various capacities and retired as Professor & Head, Dept. of Horticulture and Associate Dean (Principal), S.V. Agricultural College, Tirupati, Andhra Pradesh in 2005. The total service put in is 33 years including Ph.D. studies, of which 16 years is in research at Bapatla, Kodur and Sanga Reddy, and 17 years in teaching at Hyderabad and Tirupati. He did research mainly in citrus, mango, guava, cashew and Ber. Under teaching he taught courses in fundamentals, fruits, medicinal and aromatic plants, plantation crops at UG level and special courses in citriculture, viticulture, mango and dryland horticulture at PG level, under extension activities trained farmers of citrus, mango, guava and cashew in Andhra Pradesh for state Dept. of Horticulture, Andhra Pradesh. He is the life member of eight Professional Horticulture Societies in India. Published 60 research papers, more than 100 popular articles in Telugu, contributed chapters on citrus etc.

in package of practices for horticulture crops by Andhra Pradesh Agricultural University, Hyderabad, Andhra Pradesh and Handbook of Horticulture, ICAR, New Delhi. He co-authored with late Dr.C.B.S. Rajput book "Citriculture" published by Kalyani Publishers, Ludhiana, authored two books "Fruit growth and fruit drop in mango" and "Mango malformation disease: Etiology, Epidemiology and Management" published by Kalyani Publishers, New Delhi, as editor of "Postharvest Management of Mango (*Mangifera indica* L)" published by BS Publishers, Hyderabad, contributed a chapter "Role of plant growth regulators in mango" along with other authors in "Plant Growth Regulator in Fruit crops" edited by Dr.S.N. Ghosh et al., (2018).

Recipient of "Life Time Achievement Award" in 2006 and "Best Member" Award in 2018 of Indian Society of Citriculture by Indian Society of Citriculture, Nagpur, India.

Dr. C. Madhumathi

Born in 1967 in Andhra Pradesh. She obtained B.Sc (Ag) degree in 1990, M.Sc (Ag) Horticulture in 1992 and Ph.D in Horticulture in 2010 all from Acharya N.G.Ranga Agricultural University, Andhra Pradesh. She joined in Department of Horticulture, Andhra Pradesh as Horticulture Officer in 1993 and left in 1998 putting about 4.5 years. Next she joined as Asst. Professor in 1998 in Acharya N.G. Ranga Agricultural University, Andhra Pradesh and rised gradually to Principal Scientist (Horticulture) in the year 2016. She has put in about 19 years' service in research and 3 years in teaching and total service in the University is 22 years till to date. She served the University in various capacities and is now heading the Citrus Research Station, Petlur, Andhra Pradesh. She has handled ten research projects on vegetables and developed production technology. She specialized in fruit crops (Mango, Guava, Papaya, Banana and Citrus) vegetable crops (drumstick, cole crops and melons.) and flower crops (tuberose, gladiolus and chrysanthemum.) She assembled 5 accessions/varieties of drumstick at Kodur, Andhra Pradesh and studied their performance and claimed that PKM - 1 cultivar is most suitable for Andhra Pradesh. She disseminated production technology of above horticultural crops to farmers in several training programmes. She received two gold medals for achieving first rank in M.Sc (Ag) Horticulture. She is a recipient of "Best Achiever Award 2017 by Y.S. Parmar University of Horticulture and Forestry, Solan, Himachal Pradesh and 'Padmasri' Dr.I.V. Subbarao *Rythu Nestham*" award for outstanding research work in Horticultural crops for the year 2018. She published 25 research papers in National Journals and 26 popular articles (in Telugu). She published a book "Hints on Turmeric Cultivation" in Telugu in 2016. Presently she is working" on "Off season production in Drumstick" and "Crop regulation in Acid lime. She has contributed a few chapters in this book "Production Technology of Drumstick (*Moringa oleifera Lam.*)"

Smt. G. Jyothi

Born in December 1977 in Kurnool district, Andhra Pradesh. She obtained B.Sc.(Ag.) degree in 2000 from Acharya N. G.Ranga Agricultural University with 'Agricos Gold Medal' for securing highest OGPA at S.V. Agricultural College, Tirupati. She got II Rank in PG Common entrance test and secured highest OGPA in M.Sc (Ag) Horticulture in 2002 from Acharya N. G. Ranga Agricultural University. She worked on Moringa nutrition for her M.Sc (Ag) degree. She joined as Horticulture officer in 2002 and worked for four years in the Department of Horticulture, Anantapur, Andhra Pradesh. Later she joined as Assistant Professor in Acharya N G Ranga Agricultural University in 2006 and opted for Sri Konda Laxman Telangana State Horticulture University in 2015 and is continuing in the same capacity. Presently she is pursuing Ph.D. in Horticulture under in- service quota in the Horticulture University. So far, she has published five research papers and ten popular articles and gave six radio talks. She has put in two years' service in research and presently she is in teaching and has put in eleven years in teaching. Earlier she has contributed a chapter "Postharvest losses in mango" in Postharvest Management in Mango (*Mangifera indica* L.) edited by Dr. R. SRI HARI BABU published in 2015.

CHAPTER 1

Introduction

Moringa (*Moringa oleifera* Lamk) is considered to be one of the most useful plants on the earth, as a nature's gift to the mankind, because every part of the plant can be used, as food, feed, medicine and industrial raw material. It is indigenous to lower sub-himalayas extending from Afghanistan through Pakistan, India and Nepal (Fig. 1.1) (Fahey, 2005). Moringa belongs to the genus "*Moringa*" of family *Moringaceae*. The genus '*Moringa*' has 13 species, which are native to Indian subcontinent and Africa. Out of the 13 species, *Moringa oleifera* Lamk is well known and widely cultivated throughout the world compared to other species. Besides *M.oleifera, M.stenopetela* and *M.concanensis* are found in Indian sub-continent.

The name "Moringa" has derived from a Tamil word "Murangai". It is commonly known by several names in different languages and different regions of moringa growing. Other names of *Moringa oleifera* include drumstick (because of the typical shape of its fruit as drumstick used for beating the drum), horse radish tree (because of the typical flavour of roots that of horse radish roots), ben-oil tree (after the seed oil extracted from the seeds), miracle tree (as it provides nutritional, medicinal and industrial benefits), never die tree (since it survives even under harsh situation), mothers' best friend (as it increases the milk quantity of nursing (lactating) mothers), tree of life (as it provides several nutrients, vitamins, antioxidants and cures malnutrition in infants and young children). Moringa has a long history dating back to pre-Christian era and is known to Egyptians, Greeks, Romans and Indians (see history of moringa). Moringa is well known in India, as evident of its mention in ancient medical books/systems of medicine like Ayurveda, Charaka Samhita etc.

However, moringa has disappeared in the medieval period, but it was confined to rural areas. During the mid-20th century it was rediscovered and gained importance gradually not only in India, but also in other countries notably in Africa, where it is primarily cultivated for fodder and for combating malnutrition in children. According to

Fuglie (2001) moringa has gained popularity as a source of nutrition that can feed the needy and save lives as well. In recent times *M.oleifera* has gained a lot of popularity due to recent discovery of its usefulness to mankind with regard to nutrition and health of humans. Further, its wide ecological adaptability, low demand for soil nutrients, and water and relative ease with which it propagates through both sexual (seed) and asexual (cuttings) means, make its production and management easy, which was also considered as a reason for its popularity and spread of cultivation.

Interest in moringa in recent times is skewed towards its medicinal, pharmaceutical and nutraceutical attributes, hence much research has gone into these aspects as evidenced from volume of literature published. Moringa gained importance due to its multiferous uses. It has significant economic importance because of its several industrial uses. Moringa is a boon to the farmers which gives income particularly in poor and marginal lands. It satisfies the demand in alleviation of malnutrition. Moringa seed oil is a much sought commodity in the formulation of skin care products and delicate machinery and making biodiesel and has great demand in international markets.

Moringa is a store house of nutrients, vitamins, minerals, amino acids and antioxidants. Each and every part of the plant is edible and useful with its own nutrition, flavour and taste. It is one of the most incredible plant endowed by the divine as its nutritional and medicinal properties have immense potential to manage malnutrition, prevent and heal many human maladies numbering 300.

Moringa provides multiple uses, due to its nutrient rich parts with medicinal properties etc. It has several uses and utilization. According to Fuglie (1999) the many uses of moringa include both for humans and animals, medicinal and industrial. It is useful in agro-forestry, agri-horti-silvi programmes, in checking soil erosion, thereby conserving soil, in reclamation of mined areas and improving soil fertility through its leaf litter. The most useful contribution of moringa plant is combating malnutrition of infants in developing countries.

In view of its richness in nutrients and medicinal properties, there is great demand for moringa products and by products, which is ascending along with time. Great demand exists for its pods throughout India followed by its leaves. The demand for leaves both for human and animal consumption exists in Africa and Philippines. In Europe there is much demand for byproducts of moringa as nutrient supplements and cosmetics. Of late, the demand for seed rised much for its oil, useful in delicate machinery, and making of biodiesel and seed powder for water purification and seed cake as manure. There is much demand for seed for raising new plantations of moringa. The demand for products and by products of moringa is increasing by leaps and bounds year by year, as evidenced by the volume of trade and income out of such demand.

Several benefits have been noticed due to cultivation of moringa. It can be grown throughout the year at minimum cost. Ratooning of the crop is a boon for the moringa farmer as it avoids establishment of the crop every year like other vegetable crops. Moringa promotes environmentally sound economic development of a region. It alleviates poverty, malnutrition and nutritional deficiencies. The crop provides a life to the drought prone areas, with scanty rainfall and helps in providing beneficial land use and nutrition to humans and animals. It has a vital role in providing food security throughout the year at cheap cost.

Cultivation of moringa has great prospects because of the increasing awareness of its nutritional and medicinal values and thereof. Further, its cultivation has proven more remunerative at less cost of production. Moringa being perennial offers long life for more than 15-20 years in the same field, hence investment is comparatively less and non-recurring. People in drought prone areas are much benefited because of high returns from drumstick cultivation (see success stories). Contract farming is also possible in moringa. It is a versatile crop that can be grown as a solitary tree for vegetable purpose (pods) and seed purpose, as a perennial crop and under intensive cultivation for fodder. It is a potential crop for dryland horticulture and can be successfully grown on marginal, cultivable waste lands, degraded soils and mined areas. Moringa is a climate change adaptable crop and hence is suitable to mitigate climate change (Ndubuaku et al., 2014). Hence, its introduction into different agricultural land use systems can be beneficial to both the farmer and surrounding ecosystems (Foidl et al., 2001). It can be grown in areas with high temperature and low water availability and scant rainfall, where it is difficult to grow other agricultural / horticultural crops.

In view of the above, cultivation of moringa has bright prospects provided the Government encourages farmers, particularly in drought prone areas to start moringa cultivation by creating awareness among the farmers about its high returns, nutritional, medicinal and industrial value, through extension programmes.

However, the cultivation of moringa is beset with some problems which can be solved through research by establishing a 'Moringa Board' on the line of 'Tea Board' and 'Coffee Board'.

CHAPTER 2

History of Moringa and its Cultivation

Moringa (*Moringa oleifera* Lam), drumstick tree, horse radish tree, ben oil tree etc. the best known and widely cultivated species of genus *Moringa* of family *Moringaceae* has a long history. Historians traced moringa's history to atleast 150 BC, when ancient people in what is now Madagascar and Himalayan foot hills, used the tree's lustrous leaves for energy and improvement of health. For centuries, the moringa (drumstick) tree is in use as food and medicine. Moringa dates back to 2000 BC when it was found and described as medicinal herb, particularly in India. Historical evidences / proofs revealed that ancient Kings and Queens of Egypt, Greek and Rome used moringa leaves and fruits in their diet to maintain mental alertness and healthy skin. Moringa has a long history of success according to India's age old medical system viz., Ayurveda for centuries which used moringa as healing plant to prevent as well as treat over 300 health problems of humans.

It is recorded that in ancient India, during the era of Mauryans, the warriors were fed with an elixir made from moringa leaves in the war front, to give them strength and stamina to fight the war. The elixir drink was believed to add extra energy and relieve them from stress and pain incurred during war. It was recorded that Alexander the Great had been in a war against Mauryans of India in 326 BC. Moringa leaf elixir administered to soldiers of Mauryans gave them super human strength, required little sleep, and they never got sick and their war wounds healed rapidly and all these contributed to win the Alexander in war in the end.

Ancient Egyptians treasured moringa seed oil as a protection of their skin from the ravages of desert weather in Egypt (Doerr and Williams, 2007-09). A tomb dated from 1550-1292 BC in Egypt was found to contain 10 jars of sweet moringa oil, thought to have been used in the funeral processions of Kings and Queens. (Anon, 2014a). Lisa Mannich in "An Ancient Egyptian Herbal" gives the recipes that were used by the ancient Egyptians. It was mentioned that moringa oil was used as a carrier oil which would be mixed with other various ingredients for medicinal purpose (Anon, 2015c).

Later the Greeks found many health benefits and uses of moringa and introduced it to the Romans. The Egyptians, Greeks and Romans extracted edible oil from the seeds of moringa, used it in perfumes and skin lotions. In Qasr Ibrim, once a major city in what is now Lake Nasser, trees of *Moringa peregrina*, one of the 13 species of genus *Moringa*, fruits were found to be present as early as 7th century.

Biblical readings confirm that moringa plant has been used for millinia as medicine or as medicinal tonic (Anon, 2014). The branch that Moses used in the spring, the trees found in the garden of Eden and the trees by the river of life from the "Book of Revelation" may have contributed to the health of the population of that time. Some even believe that the Bible references of the moringa tree in the "Book of Revelation was the "tree of life" and the book of Exodus as a tree that helped make the bitter water sweet. Perhaps moringa is the tree in the garden of EDEN referred in the Bible, as it sure qualifies to be the one-tree. References in the Bible, when Moses led fleeing Israilities to the village of Marah, they could not drink the water as it was bitter and when Moses asked God what should be done, God showed Moses a tree which when Moses had cast it into water, bitterness disappeared and turned to sweet water (Exodus, 15, 25).

There is also mention of moringa in "Quaran" which has sometimes been thought of as the "Olive tree" but, modern thinking goes more towards the "Moringa tree" as the seeds of moringa contain 40% oil, whereas the olive seeds only 20% oil (Anon, 2015c).

Thus, moringa tree had been around for thousands of years. Yet we are just starting to realize the importance and power it contains. It is often wondered could this tree be the key to ending poverty, and malnutrition of the world. The tree is now re-discovered in many areas of the world having a significant role in the above two aspects. At present moringa is being used around the world to stop starvation and malnutrition in humans. Recently, there has been a surge in moringa popularity as its leaves are harvested fresh in India and ground into powder which was then stored in air tight containers for marketing domestically and internationally.

Moringa was originally grown in India by the native Dravidians and later by Aryans in each and every home yards. It disappeared in medieval times. It was neglected and forgotten as a medicinal tree, though it continued as a vegetable tree. During the colonial times, it was rediscovered as a horse radish tree by the Britishers as a replacement of horse radish plant. It's significance as a cure against malnutrition in children has been realized in India, Africa, Philippines etc and boosted its cultivation. Five decades ago protein malnutrition with vitamin-A deficiency was rampant among pre-school children (<5 years) in India. Nutritionists tried to identify an inexpensive supplementary food to overcome this problem and identified moringa leaves as the solution for malnutrition in children. Then drumstick seeds and limb cuttings were distributed in villages and urban slums for growing in the back yards. Mothers were taught easy recipes. It has paid dividends (Anon, 2015c).

History of Moringa Cultivation

According to Odee (1998) the history of moringa cultivation dates back to several ages. Historically there is evidence that cultivation of *Moringa olifera* Lamk started in Indian sub-continent, the native home of moringa and dates back to thousands of years. According to Sushruta Samhita and traditional Ayurvedic medical texts moringa was grown to heal or prevent hundreds of ailments and diseases of humans (Beaulah et al., 2010). Moringa was grown in India first by Dravidians and later by Aryans in each and every home yards to serve both as vegetable and medicine (Beaulah et al., 2010). In Indian sub-continent moringa has long been grown / cultivated for its edible leaves, pods and seeds to extract oil and prepare flocculant with dried seed powder and for making leaf powder.

Though drumstick originated in the North West sub-himalayan foot hills, it did not gain importance and cultivation in North India, but gained foot hold in South India.

Drumstick is known and grown in the Indian sub-continent since long. However, all these long years it was/is confined to home steads, solitary trees at animal sheds, wells, field or tank bunds and group of trees in waste lands of rural areas and as isolated plants in hedges and fences. It was never cultivated on garden scale till 1980s. Till that time only unnamed local perennial types or types with local names (confined to each local area) were grown. With the demand of bulk quantities of moringa, farmers started increasing the number of trees per unit area through branch cuttings of perennial types in late 1980, but were not successful, their fruiting being eratic. However, after the introduction of a perennial type "Jaffna" from SriLanka farmers evinced some interest in the cultivation of drumstick on garden scale in Tamil Nadu and extended to other South Indian states on a small scale. In early 1990's farmers of Tamil Nadu started growing perennial types as an inter-crop on field scale and their allies were cropped with vegetables and sorghum. This system was evolved as moringa offered some protection to alley crops from drying winds during summer season and from soil borne insect pests and disease and enriched the soil fertility by its litter and provided some additional income to the farmers through the sale of pods. Next, farmers found that growing moringa crop during summer season was more remunerative though it is one of the cheapest vegetable during summer (Beaulah et al., 2010). However, such cultivation of perennial moringa did not progress much. Due to the inherent demerits like long gestation period (3-4 years), difficulty in propagation, oozing of gum always from the trunk, attack of swarms of hairy caterpillars on the foliage and defoliating trend and their appearance in groups on the trunk, often entering the houses, heavy incidence of fruit fly, short pods, with thin flesh, cultivation of perennials including introduced "Jaffna" did not make much head way in area expansion under drumstick.

Commercial cultivation of moringa started picking up after the introduction of annual types viz., PKM-1 and PKM-2 by Tamil Nadu Agricultural University in 1990's. Since then the area and production of moringa are increasing year by year in India, which stands top in area and production of moringa pods globally. Annual types have some merits over perennial types. Annual types are precocious – come to fruiting by 6th month after seed sowing, can be propagated easily by seed and not susceptible to hairy caterpillar and produce long, fleshy and tasty pods, compared to perennial types. Due to universal adaptability to various agro-climatic situations of annual types, commercial cultivation of drumstick gained momentum in India especially in Andhra Pradesh, Telangana, Tamil Nadu, Karnataka, Kerala, West Bengal, Odisha, Gujarat and Maharashtra.

PKM-1is seed propagated and precocious and hence spread fast across the Southern States of India to begin with and later to many of the African countries and tropics of entire globe with in a span of five years in as much as the moringa was accepted as a crop to remove malnutrition of humans in the world.

The economic analysis of moringa cultivation has illustrated that commercial cultivation of moringa could be a profitable venture to the farmers and could alleviate their poverty in a short time (Sherker, 1993). Further the ease with which it is cultivated and ready marketing and export possibilities, prompted and encouraged farmers to take up cultivation of moringa. Due to the above reasons, the interest in drumstick cultivation and area under its cultivation increased many folds during the last 2 to 3 decades in India.

Because of its unique flavour, aroma and taste moringa became popular in entire South India, where any meal without moringa, pulses and curd is considered incomplete. So the demand for moringa pods also increased due to urban settlements and migration of people to urban areas particularly from south to north and also to other countries. With the taste and flavour as deep rooted in them, ethnic Indian population settled elsewhere in the world, predominantly in Gulf countries, America etc, longed moringa in their diet. All these simultaneously led to the cultivation of moringa on commercial scale, which is ever increasing. Thus, moringa gained foothold as a commercial crop particularly in Southern States (Beaulah et al., 2010).

The history of cultivation of drumstick in countries other than India is not known and not available in the scanned literature. In several countries the cultivation of *Moringa oleifera* Lam is recent one only after the introduction of the crop, as a remedy to child malnutrition by NGO's. Unlike India, however, drumstick is cultivated mainly for leaves to fight child malnutrition and lactating mothers and for fodder purpose in Africa and Philippines. Now moringa plantation for seeds (for extraction of oil, biodiesel etc) is being promoted at global level.

Thus, moringa started its journey as an individual tree, as intercrop and finally ended in commercial cultivation for leaves, pods and seeds all over the world. Further, the cultivation of moringa started with perennial types and continued with annual types, because of certain disadvantages of perennial types (see above).

CHAPTER 3

Prospects and Constraints of Moringa Cultivation

Prospects

Moringa / drumstick (*Moringa oleifera* Lam) is a versatile crop, acclaimed as a multipurpose crop. It is considered to be a "super food" a store house of nutrients, vitamins, antioxidants etc and claimed to cure 300 ailments of humans as per ancient "Ayurveda". Its cultivation is highly profitable, remunerative with less investment. It alleviates poverty of poor rural people within a short time of 6 to 12 months. However, till second half of 20th century it was neglected and was an underutilized crop. Its popularity is increasing day by day because of its nutritive richness, medicinal and industrial uses and its ability to adapt to various environments. It is well suitable for kitchen garden, field bunds, near cattle sheds; fits well in different cropping systems etc. In view of the above, moringa has great prospects for cultivation in future:

1. Moringa's cultivation can be extended to arid and semi-arid areas as a dry land horticulture crop and can successfully be grown on marginal and cultivable wastes; with high temperature and less water availability, where it is difficult to grow any other crop.
2. Its cultivation is a profitable proposition in drought prone areas, for alleviation of poverty.
3. It is a versatile plant that can be grown as a tree or as perennial fodder plants under intensive cultivation for biomass production.
4. It does not need much manuring and fertilization and irrigation and is not cost intensive, hence its cultivation can be taken up by small and marginal farmers with advantage.
5. It is a fast growing and precocious plant giving yield within a year and is amicable for ratooning and can stay in the field for 3-4 years at a stretch (like mango, citrus, guava etc.) with non-recurring establishment costs every year.

6. It can benefit humans nutritionally and medicinally as energy source and economically

7. Contract farming is also possible with moringa farming.

8. In India it is mainly cultivated for its fruits (pods) and less for leaves, which are mainly used for culinary purpose. However, as leaves can be used for making leaf powder etc and can also be fed to cattle (fodder) moringa cultivation can also be encouraged for making by products and as cattle feed.

9. It is a climate change adaptable crop, hence it can be grown in areas of climatic changes (Ndubuaku et al., 2014b).

10. The relative ease with which it propagates through both sexual and asexual means and its low demand for soil nutrients, fertilization and less water, make its cultivation easy and less costly. It can fit in many cropping patterns. Hence, its introduction into agricultural land use systems is beneficial to both farmers and ecosystem (Foidl et al., 2001).

Thus, moringa's cultivation has bright prospects in the country, provided the following constraints (problems) are addressed and solved and backed by government intervention and encouragement.

Constraints (Problems)

In spite of high demand of products and by-products of moringa allover the world and bright prospects in future, the cultivation of moringa has not progressed much, because its cultivation is beset with several constraints (problems), which differ from region to region in the country. Despite the fact that much research has undergone, which resulted in the development of standard cultivation practices, still moringa faces following problems:

1. Lack of region wise high yielding variety(s)

2. Lack of suitable varieties exclusively for biomass (leaf) production.

3. Lack of suitable varieties for subtropical conditions of North India, to extend the moringa cultivation in North India.

4. Non-availability / limited availability of good quality planting material especially in annual moringa which are exclusively propagated by seed, as seed is being utilized for other purposes.

5. Lack of suitable pruning methods for ratoon crop management.

6. Erractic / irregular flowering behaviour and shy bearing in certain states like Andhra Pradesh and Telangana
7. Lack of suitable off-season production technology
8. Lack of post-harvest management to avoid post-harvest losses for maintaining sustainable prices of the products
9. Lack of cold-storage facilities in production centres
10. Lack of establishment of value added product centres to export to overseas markets.
11. Incidence of pests and diseases and back of their fool proof management.
12. Lack of organized marketing system and cooperative marketing system and existence of market problems.
13. Lack of standard cultural practices for organic production of moringa. Lack of encouragement and support for enhancing organic production of moringa.
14. Lack of nodal agency (like coffee / tea / rubber boards) to conduct research on all aspects of moringa cultivation and extension to take up the research results to the farmers.
15. Lack of awareness among the population about the importance and benefits of moringa and its products and by-products in human health and about industrial uses.
16. Lack of government support and investment in moringa research and cultivation.

CHAPTER 4

Origin, Spread and Distribution

Moringa oleifera Lam has originated in north-west lower sub-Himalayas through Afghanistan, Pakistan, India and Nepal (Fig. 4.1) (Fahey, 2005) as back as 5000 years ago (Umbertor 2000). It has later spread into the southern parts of India, where it gained permanent foot hold. The cultivation of moringa tree in due course has spread to most of the countries in tropics and sub-tropics in all the five continents, to most parts of Asia, Africa, South America, Central America, Southern parts of North America, Australia and some parts of Europe (Baba et al., 2015) (Fig. 4.1).

Arab traders and European Christian missionaries are instrumental in the spread of moringa worldwide. It was introduced into Nigeria, by Arab traders as indicated by Arabic names, which is now being grown in most parts of Nigeria, where it acquired many local names (Auwalu, 2009). *M.oleifera* might have been introduced into Zimbabwe during European occupation by Europeans or probably long before by Arab traders (Palgrave, 1983). However, it is believed that the moringa tree was introduced into Zimbabwe by Indian traders arrived in quest of gold and ivory. It was introduced into Brazil in 1985 (Chaves et al., 2005). Thus, it has spread throughout India, Pakistan, Asia Minor, Arabia, South America and Africa. It is now found in many regions of the world like, the Pacific Islands, Central America (Mexico), Caribbean Islands and North America (Morton, 1991; Mughal et al., 1999).

According to Nadkarin (1976) and Ramachandran et al., (1980), *M. oleifera* is widely distributed in sub-Himalayan region of India (Fig. 4.2), Sri Lanka, North East and South West Africa, Madagascar and Arabia, where it is naturalized. It is widely distributed in more than 100 countries all over the globe.

It is widely cultivated in all the continents wherever it was spread and distributed (Ramachandran et al., 1980; Morton, 1991). Currently, it is widely cultivated in more countries, notably in India, Pakistan, the

Philippines, different African countries, and in some parts of North America, Central and South America, and Europe and Australia and Pacific Islands (Fig. 4.3).

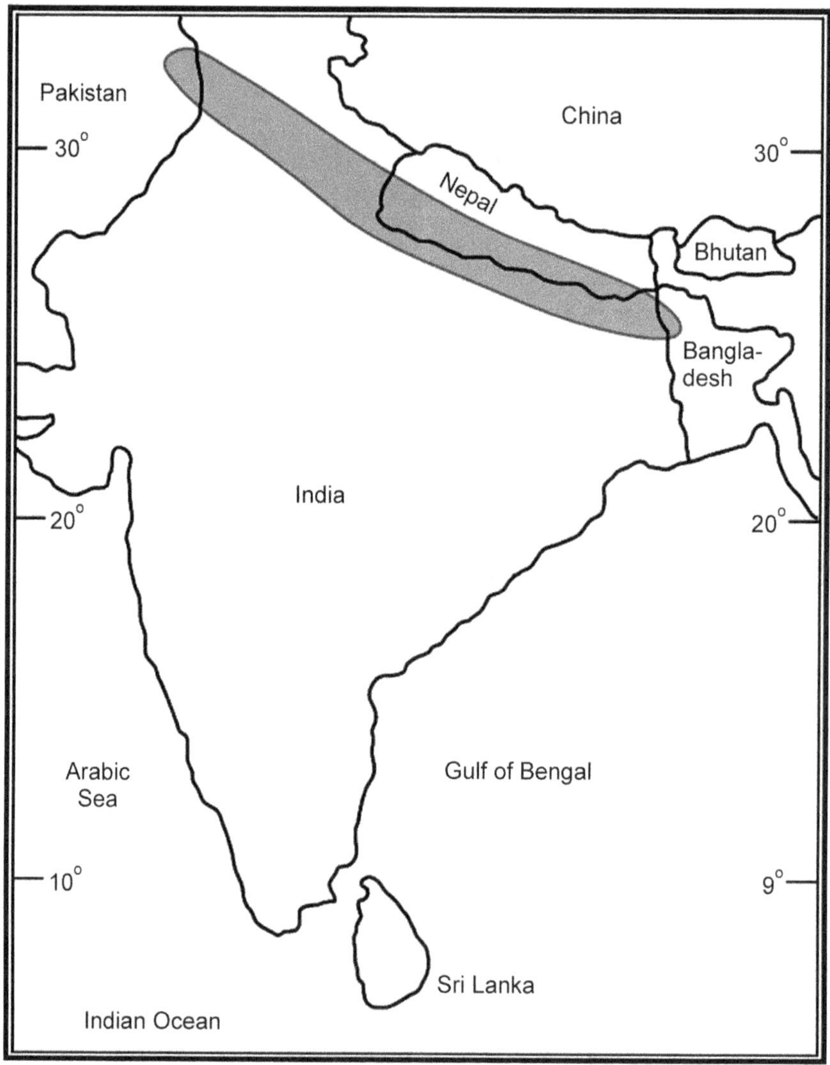

Fig. 4.1 Map showing the origin of Moringa oleifera.

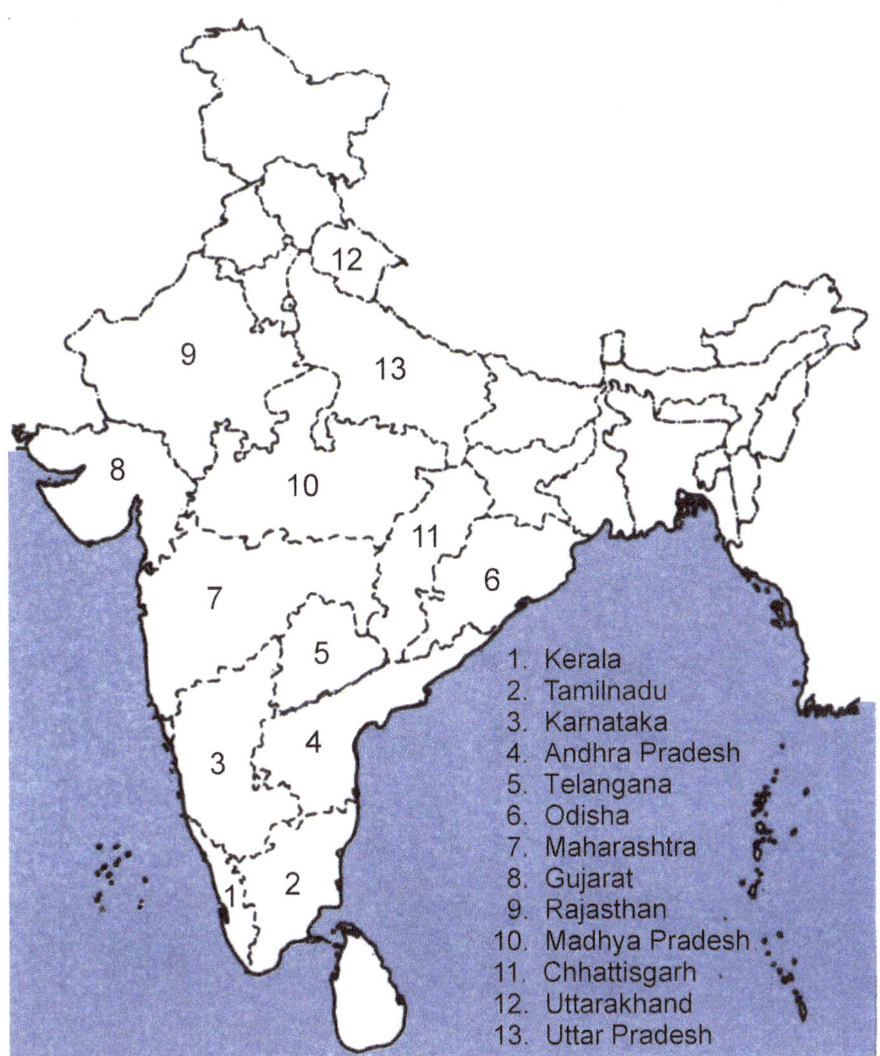

Fig. 4.2 Distribution of drumstick in India.

Fig. 4.3 Distribution of Moringa in the world

CHAPTER 5

Climate

Climate decides whether a particular plant / crop is suitable for a particular area. Climatic variables (elevation, temperature, rainfall etc.,) have strong bearing on the adaption, survival, growth performance and quality of its produce etc. of a given crop. Therefore, before venturing to take up cultivation of crops especially commercially, the farmer should have adequate knowledge and understanding of the climatic requirements of the given crop and the prevailing climate of the proposed areas, its suitability to the crop in question and the influence of climatic variables on the crop.

Drumstick (*Moringa oleifera*) with regard to climate is an extremely adaptable species. It grows well throughout the warmer zones ranging from tropical to sub-tropical zones, and from dry savannahs to rich forest soils in Africa (Anon, 2006). Moringa has large ecological elasticity and adapts well to the most diverse weather conditions. It tolerates wide range of environmental conditions. Moringa adapts to both dry and humid tropics as well as wet forests, from arid to humid climate with a prospect of growing in a wide range of land use systems (Paliwal et al., 2011b). It is predominantly a crop of dry and arid tracts and is a plant of low land tropics and hot dry regions (Somali et al., 1984). It has adapted to various ecosystems and farming systems wherever it was introduced. For example it was well adapted to the climate and environmental conditions (mild winter and warmer summer) of South Taiwan with in twenty years of its introduction (Palada et al., 2007).

Moringa thrives best under tropical and sub-tropical climates in India (Verma and Chaarasia, 1993). One of the reasons that the moringa can thrive well in arid zones is that it possesses a long tap root and extensive deep root system, which also make it valuable against soil erosion.

Moringa is particularly suitable for dry region as it can grow using rain water without irrigation (Paliwal et al., 2011a). Hot and humid climate is suitable for growth and dry climate for flowering and fruiting.

It is known to attain best growth and perform well in tropical insular climate in the plains of South India. Moringa grows best in the hot semi-arid tropics with hot to warm climate. Under these conditions it produces leaves and pods all the year round. According to Verma and Chaarasia (1993), moringa grows well from Tarai region of Uttarakhand (India) to coastal areas of Southern India. It can be grown in arid zones with little water and in dry climates or in climates with long period of dryness and shorter periods of rains.

Moringa is ever-green in tropics and deciduous in sub-tropics (Muhl et al., 2011b). In North India, the leaves are shed in December-January and the tree enters into dormancy and puts forth new foliage during ensuing February – March months followed by flowering and fruiting (Benthal, 1946). However, during severe winter even in the tropics the perennial types of moringa undergo dormancy by shedding leaves and once the temperature rises in spring, such plants put forth new foliage (Seemanthini, 1964).

Moringa is cold tender, hence not recommended for growing in countries with cold climate especially in open conditions. However, it can be grown indoors or / and green houses. Under open conditions, moringa plants should be protected from cold temperatures by mulching the tree basins to protect the root system. Well mulched trees can survive severe frost. The tree may freeze down to the ground level, but survives and will grow new shoots in spring. Young trees are more susceptible to cold climate, while the grown up trees show resistance to winter temperatures by entering into dormancy.

Elevation (Altitude)

Moringa is a tropical low land plant (Radovich, 2009). It is found on either side of equator between 23⁰ N and S latitudes and at elevations of 800 to 1200 metres above mean sea level. But normal distribution of moringa is from sea level to 500 metres, but found to occur at still higher levels, with a change in its behaviour. When grown at 1600 m it fails to flower (Jahn, 1991). However, it manages to survive at 1000 m (as in Mexico). At 1500 m elevation, it is a complete failure. Hence, it was concluded that it can grow from sea level to 500 m (Radovich, 2009). Ideal elevation is therefore, below 600 m for *M. oleifera* (Beaulah et al., 2010; Majhi, 2013). Originally, moringa was recommended for planting only at altitudes below 600 m in upland areas of Africa. However, the discovery of healthy stands at elevations 1200 m in Mexico and over 2000 m in Zimbabwe, demonstrated that it is much more adaptable than supposed (Anon, 2006). But the best elevation for moringa is from sea level to 500 m for commercial cultivation.

Temperature

Atmospheric temperature is the key climatic variable, which greatly influences the growth, development and various physiological processes in plants. Plants perform satisfactorily with in a range of temperature and lower or higher temperatures than the optimum range of temperature are unsuitable as they cause some damage and disruption of plant processes. Further, lower or higher temperatures determine the choice of location of growing a particular plant /crop.

Growth of *M.oleifera* is evidently favoured by high temperatures of more than 25ºC and it puts forth luxuriant growth at a temperature range of 25º to 30ºC. At Panthnagar (Uttarakhand, India), moringa was found to tolerate a maximum temperature of 38º-40ºC and a minimum temperature of 1º-5ºC (Tewari et al., 2003). Moringa survives a temperature range of 25º-40ºC, but tolerates upto 48ºC in shade (Palada and Chang, 2003; Dhaker et al., 2011). Nigam and Mishra, (2003) reported 25ºC as optimum for *M. oleifera* under Uttarkhand conditions.

For optimum leaf and fruit production *M.oleifera* requires a high daily temperature of 25⁰-30⁰C (77-86ºF). When the temperature falls below 20ºC (68ºF), plant growth is slow (Majhi, 2003; Paliwal et al., 2011b). Higher temperatures than 40ºC result in heavy flower shedding (Veeraraghavathathum et al., 1996; Kathiresan et al., 1999; Thamburaju and Natarajan, 2010). Moringa, however, does not get killed by very high temperature and does not even drop leaves, provided there is adequate water in the soil for uptake.

Frost

Moringa oleifera is frost hardy, it tolerates light frost (Beaulah et al., 2010), but gets injured by severe frost. It cannot tolerate higher cold and snow-fall although its origin is sub-himalayas. It survives mild freezing temperature in winter. Light frost freezes the tree to the ground level (Beaulah et al., 2010).

In areas where the temperature drops below freezing as in some parts of California and Texas in USA, trees will still survive in which case the tops of trees freeze but deep roots will survive and as soon as weather warms up in spring the trees will shoot out new shoots within a couple of months and start giving leaves for vegetable purpose. In areas where the soil freezes to a depth of six feet the tree may be cut down to about 8" stump, insulate the stump with leaf mulch, so that the soil of 2' radius of the trunk won't freeze. When the snow melts and temperature falls below 40ºF insulate with mulch and wet trees beneath. This result in the faster emergence of shoots and leaves will be available for harvest within two months. In areas of severe cold, flowering is delayed. Continuous frost and sub-zero temperature will kill the trees. In areas with light frost (but without sub-zero temperatures) moringa can be cultivated with adequate care and protection.

Rainfall

Moringa is a crop of low rainfall areas, thus, is mainly grown in arid and semi-arid regions of India. It is tolerant to annual precipitation of 250-1500 mm. And grows well where precipitation is between 260 and

2150 mm with access to ground water. According to Palada (1996) moringa tolerates an annual rainfall of 760 to 2250 mm in Taiwan. It is extremely vigorous and grows luxuriantly in dry climates with long periods of dryness or short periods of rainfall. Rain forests in tropics which receive higher rainfall are not suitable as moringa is intolerant to wet soil. Maximum rainfall was estimated as 3000 mm with a minimum of 550 mm (Doerr and William, 2007-09). Very high rainfall during flower initiation affects it and as a result the plants will remain in vegetative phase (Beaulah et al., 2010). High rainfall during blooming also causes flower shedding resulting in less or no fruit set and yield.

Distribution of rainfall evenly is much important than the quantum of rainfall. If rainfall is continuous throughout the year the trees exhibit continuous flowering and fruiting (Beaulah et al., 2010). Therefore, areas with well distribution of rainfall and intermittent dry spells are suitable for growing moringa. If the soil is in a wet area or that receives abundant rain creating wet soil with poor drainage planting of moringa on hill slopes or mounds can help encourage run off of rainwater and prevent excess wetness in the soil and contact of water with tree trunk.

Water logging

Moringa plants are very sensitive to water logging in the soil (due to poor drainage or excess rainfall or heavy clay soils) and typically will not survive under prolonged flood or water – logged conditions. In a field trial, most of the moringa accessions from different countries survived the intermittent flooded conditions of the field. It was observed that moringa plants tolerated waterlogged conditions for several days and when rains ceased, water receded from the fields and soil moisture decreased, dryness returned, plant got rejuvenated with new flush of leaves. In such conditions varieties that tolerate prolonged water logging conditions should be planted. In water logged soil the roots show a tendency to rot.

Drought

Moringa is drought tolerant and exhibits certain degree of Xerophytic behaviour but yields less foliage under water stress conditions particularly where rainfall is below 300 mm per annum. Such water-stressed trees, however, recover with the arrival of rains (von Madell, 1986). Its deciduous nature during the dry period and its enlarged underground parts make moringa very drought tolerant (HDRA, 2002). In drought prone areas moringa requires relatively high water table to be productive.

Being drought tolerant, moringa is highly suitable for water shed development programmes. It is, especially adapted to dryness and may resist several months of drought, due to its swollen roots which store water (Anon, 2006).

Sun

Moringa is a sun loving plant (Paliwal et al., 2011b). It requires higher solar radiation and grows best in direct sunlight. It requires adequate sunlight for production, in the absence of which trees fail to produce leaves and fruits. In South India during winter (October-January) the light intensity as well as temperature will be less, which results, in very less flower induction and fruitset and this period is considered as a lean period for moringa (Beaulah et al., 2010).

Although moringa is a tropical plant, the tree could be damaged by excessive heat especially in full sun, that might scorch the leaves. In areas subject to high heat during summer months, it is advisable to locate the trees / plantation in a place that is not exposed to sun but to shade during the hotter after noon hours. Under these conditions moringa trees can survive without significant damage. Alternately, keeping two or three layers of organic mulch around the base of the trees can also help to protect it from heat by keeping the soil and roots cool (Marie).

Wind

Moringa is a fragile tree, but is wind tolerant, hence it is sometimes used as a fence and wind break. Excessive windy conditions break its branches and also cause the trees to dry out in South India (Beaulah et al., 2010).

Finally, it can be concluded that moringa for optimum leaf and pod production requires dry tropical and sub-tropical, arid and semi-arid regions with high average temperature range of 25^0-30^0C without frost / freezing, well distributed rainfall (1000-2000mm), high solar radiation and wind less weather, well drained neutral (6.0-7.5 pH) open soils (Odee, 1998; Radovich, 2009; Beaulah et al., 2010; Paliwal et al., 2011b). The right climates for moringa are found throughout the dry tropical and sub-tropical areas of the world.

CHAPTER 6

Soils

Though moringa has adapted to a wide range of soils and soil conditions, it does not grow equally well in all types of soils and soil conditions. There is no uniformity among the scientists and farmers with regard to the soil requirements of moringa, except in respect of soil drainage, water logging and salinity.

Before taking up of commercial cultivation of moringa, farmer should have knowledge about the soil requirement and the reaction of moringa to different types of soils and associated soil conditions, when grown in different types of soils and soil conditions, for higher performance and yields.

Moringa can be grown even under harshest and driest soils, where barely any other plant can grow. This is because it tolerates a wide range of soils, pH and salinity. In fact one of the nick names of moringa is "Never Die" tree due to its incredible ability to survive, grow and produce in harsh, weathered soils. Since it is drought resistant, it can be grown even in poor soils of coastal sands (Muragavel, 2010) and waste lands (Padmavati and Manohara Rao, 2005). For instance, trials in Vietnam have shown that moringa could be grown well in hilly area in weathered soils of low fertility (Manh et al., 2003). Marginal soils with ample fertility are suitable for moringa (Nouman et al., 2014). Rain forest soils though moist are also suitable (Peter, 1978). Moringa hates waterlagged soils, but loves well-drained soils that allow free circulation of air around its roots. Gravel soils showing fast drainage are preferable for moringa. Moringa performs poorly in the best soils if they are water logged due to poor drainage. It readily adapts stream banks and savannah areas, where soils are well drained and water table remains high all the year round.

Type of Soil

Though moringa is tolerant to various types of soils (Odee, 1998) the best soils will be sandy loams free from water logging. In dry shallow soils along with hill slopes, it grows and develops poorly, hence should be avoided (Tewari et al., 2003). It grows best in sandy and alluvial soils (Palada and Chang, 2003). It prefers recently alluviated soils that have deposits of sediment and nutrients by inundation of a river or stream creating a fertile soil composure (Parrota 2009). According to Ramachandran et al., (1980) moringa grows well in almost all types of soils, except stiff clay, calcareous and water logged soils. In water logged soils the roots show decay. Therefore, in high rainfall areas and ill-drained soils, the trees need to be planted on small mounds to encourage water run-off and to avoid contact of roots with water. Moringa tolerates well-drained clay soils (Dhaker et al., 2011). It can survive droughts in clays as they have water-holdilng trend. However, in spite of this raising of moringa in clay soils should be discouraged, because of the chances of poor drainage and water stagnation. Many scientists shun the use of calcareous soils for growing moringa. The presence of bicarbomates in calcareous soils cause nutrient imbalances and nutrient deficiencies, which limit the growth of moringa and reduce production (Kishchiuk, 2000). However, Abdel Wanise et al., (2017) opined in Egypt that moringa can also be grown in calcareous soils where production can be achieved and sustainable by foliar application of nitrogen and potassium.

Moringa grows satisfactorily in marginal soil as it possesses a hardiness that enables it to survive prolonged droughts. It can with stand moisture stress and low soil fertility (Singh et al., 2014). However, low moisture in the soil leads to poor yield in spite of high fruit set due to increased shedding of fruitlets, which can be avoided by frequent irrigations during fruit set and fruit growth (Sanjay Singh, 2010).

Moringa does not grow well in used up sandy soils that are poor in fertility. In case it is necessary to use such soils for moringa, add sand with a mixture of any soil available at hand in the gardens or from outside and coconut fibre. This enables the moringa roots to penetrate deep and also water drainage. Moringa can be planted in degraded soils for soil conservation, hence it is often recommended for erosion control and reclamation of mined soils (Tewari et al., 2003).

Moringa grows luxuriantly in the sandy, sandy loams and equally well in humus rich soils (Peter, 1978). Its performance is best in dry sandy soils (Duke, 1978) including coastal soils, even though such soils are poor in fertility (Majhi, 2013). It grows very well in deep sandy loam soils, (Kathiresan et al., 1999), hence its cultivation is more or less confined to sandy soils of coastal areas in the country (Seemanthini, 1964). Sandy loams with leaf mould and peat moss are also suitable (Verma and Chaarasia, 1993). In trials in Egypt, moringa plants cultivated in sandy soils showed better response in terms of growth and pod yield compared to those grown in calcareous soils (Abdel Wanise, et al., 2017). Sandy loams containing good amount of lime are said to be best and preferable for moringa (Muthuswamy, 1954). Light soils are preferable for rooting of limb cuttings in the main fields.

The suitable soils for growing moringa, thus would be fertile sandy loams, humus rich soils (Anitha et al., 2013) well drained clay soils (Lalita Kameswari, 2004) and sandy red and black soils (Nigam and Mishra, 2003).

Soil pH

Soil reaction (pH) is one of the soil associated factors that determines the adaptation, survival, growth and performance of plants. The plants survive, grow, develop and yield well within a range of soil pH. Moringa tolerates a wide range of soil pH from 4.5 to 9.0 and grows well in alkaline conditions upto a pH of 9.0. But for commercial cultivation neutral soil pH is preferred (Beaulah et al., 2010). Slightly alkaline clay and sandy loam soils were considered the best for moringa due to their good drainage (Ramachandran et al., 1980; Abdul, 2007). However, ideal and suitable pH is 6.5 to 7.5 (Kathiresan et al., 1999). In Vietnam, moringa grows well with high biomass production in acid sulphate soils (Manh et al., 2003). It can also be cultivated at higher soil pH of 8.0 to 8.5 as in Pacific regions, where it is doing well (Beaulah et al., 2010). Moringa is growing well in calcareous soils and coastal sands with pH more than 8.0 in Andhra Pradesh (Reddy et al., 2010). Among the different varieties, PKM-1 was found to be tolerant to soil pH of 8.5, hence it can be preferred for growing in soils with pH more than 8.0 (Kathiresan et al., 1999).

Soil Salinity / Sodicity

Moringa can be grown in soils with slight salinity. However, it is believed in general that moringa tolerates soil salinity / sodicity to some extent. According to Valia et al., (1993 a, b) moringa is medium tolerant to salinity / sodicity hence, there is possibility of growing in salt affected soils, where in cultivation of agricultural crops is not economical. The growth and performance of moringa depend on the level of salinity. In pot culture studies growth, yield and physiological parameters decreased positively with increase in exchangeable sodium percentage (ESP) levels. Yield was reduced, at higher level of ESP (both fresh and dry weight). Moringa seedlings when grown in normal or artificial sodic soils showed significant decrease in plant growth at ESP 41.0. As the ESP increased from 41.0 onwards growth decreased. On the other hand, control plants performed better by 3-5 times. The growth of plants, chlorophyll and transpiration of moringa plants showed progressive decrease with increase in ESP levels in growth medium (Valia et al., 1993a, b).

Besides soil, irrigation water and water in the soil profile may be saline, affecting the performance of plants. It was reported that moringa could tolerate water electrical conductivity (EC) of 3 dSm^{-1} during germination phase, while at later stages its resistance to saline water increased (Oliveira et al., 2009).

CHAPTER 7

Varieties

Like in any other agricultural / horticultural crops, some varieties were identified, evolved / developed in *Moringa oleifera* Lam. which differ in their growth habit, leaf, flower and pod characters (Palada and Chang, 2003). Worldwide there is considerable variation in *M. oleifera* in growth pattern, rate, branching habit, time of flowering, floral biology, leaf size and colour; length, colour and other characteristics of pods in moringa. Indian varieties were principally selected and bred, keeping in view the pod characteristics viz., length, tenderness, freedom from fibre and bitterness and green colour and pulp quality and quantity.

Incidentally some of the Indian varieties released for pod purpose are found to provide foliage for human consumption (as green vegetable) for fodder for cattle, and for making dry leaf powder and other byproducts. As such there are no varieties exclusively for leaf production among the Indian varieties. However, in Dutch Indies (in India) it was reported that there are forms which flower and produce pods rarely and they are cultivated for their foliage alone (Muthuswamy, 1954). Of late, one or two varieties were evolved exclusively for seeds, for oil extraction, to be used for getting biodiesel and for other industrial applications.

The existing varieties of *M.oleifera* were developed by public and private sectors and some even by moringa farmers.

Types of Moringa

The several of moringa (*M. oleifera*) varieties are broadly classified into two groups (type) viz., perennials and annuals based on their earliness of flowering after sowing of seeds / after planting seedlings in the mainfield.

Perennial Types

These types (eco types) take some years (2-4 years) to come to flowering and fruiting on sexual propagation. Generally they do not flower in the same year of sowing / planting. They have been in cultivation from thousands of years until the advent of annual types. In India, the west and northern parts had perennial types predominantly due to which commercial cultivation of drumstick remained at low level in these regions. Similarly, in the past in southern states of India, particularly in Tamil Nadu, the cradle for the existence of several ecotypes of *M.oleifera*, only perennial ecotypes were under cultivation (Beaulah et al., 2010).

Perennials are typically propagated asexually by stem cuttings which hindered the rapid area expansion attempts, as they respond only to vegetative propagation and long distance transportation of cuttings was not possible, besides lack of adequate number of cuttings. With the demand for bulk quantities of moringa pods, farmers started increasing the density of trees per unit area through branch cuttings of perennial types in late 1980s, but were unsuccessful as their fruiting was erratic (Beaulah et al., 2010). Attempts to seed propagate these perennial types were also unsuccessful as the seedlings took some (2-4) years to flower and fruit. Further, seedlings of perennial were not true to type and had shown variation in growth and pod characters and because of cross pollination nature, the off-springs exhibited a mosaic of plant characters.

Perennials bear only edible pods, whereas annual types produce both edible and non-edible bitter pods. Perennials are typically propagated by asexual (vegetative) method through stem cuttings. Perennials yield less and generate income only after 5-6 years, hence not planted commercially.

Perennials have some undesirable features like, large trunk (1-2 metres), growing very tall (5-6 metres making harvest difficult), continuous oozing of gum from stem and attack of swarms of hairy caterpillar. Further, perennials are beset with some productive constraints, relatively long pre-bearing period, non-availability of planting material (stem cuttings), requirement of greater number of rainy days in dry regions, where water is scarce; vulnerability to pests and diseases (less resistance).

Due to the above, commercial cultivation of perennials was restricted, which favoured the development and cultivation of annual moringa varieties since 1990s.

Examples Important Perennial Ecotypes

Moolanur Moringa, Valayapatti moringa, Charakachar moringa, Chem murangai, Kettu murangi, Kodikkal murangai, Pal murangai, Puna murangai, Jaffna,Pala murangai.

Annual Types

To overcome the production constraints and the undesirable features of the perennial types of moringa since 1980s, a search was made to identity better types of moringa from the existing germplasm, available in the country especially in Tamil Nadu as it has varied genotypes, from diversified geographical areas and introductions from Sri Lanka and also to meet the ever increasing demand for pods. The search was successful and resulted in identification of some early flowering ecotypes which are sometimes called "Annual types", because they produce vegetable pods for market within a year and may be removed and new plantations are planted in the field (Kader and Shanmugavelu, 1982). However, due to precocity in flowering and bearing and persistent flowering and fruit production without any definite peak the trees become exhausted and appear to be senescent at the end of one year and hence the candidate was treated as an "annual". Except this set back in vigour, the trees donot die out on their own accord, hence it would be a misnomer to call it "annual drumstick" (Kader and Shanmugavelu 1982). However, in spite of the above objection, all these ecotypes which show precocity of flowering within a year of sowing are referred to as "annual types". In fact they are also perennials in the sense they live more than two years if allowed.

Released annual types are the products of plant breeding research conducted at Periakulam, Tamil Nadu and have now replaced most of the perennial varieties that previously dominated the commercial production in India. Main contributing factors for increase in the area under annual moringa are (Beaulah et al., 2010) :

(i)   Freedom from hairy caterpillar

(ii)   Absence of oozing of gum from trees

(iii)   Ease of propagation (by seed)

(iv)   Precocity in flowering and fruiting

(v)   Higher productivity and production

(vi)   Tasty pods with high flesh (pulp)

(vii)   Non-fibrous, non-bitter pods unlike in perennials.

However, annual types also have certain demerits (Punitha et al., 2019).

(i)   Short life span

(ii)   Requirement of more frequent replanting and

(iii)   Reduced genetic diversity and uniformity

As annuals these may have to be replanted often, hence they may be suited for areas with short growing period and climate that are too cold for the trees to survive through winters. In India, these varieties can be in production for 4 or 5 years at a stretch before replacement is required.

Ex: KM-1, Dhanraj, Bhagya, PKM-1 and PKM-2 of which, PKM-1 and PKM-2 are the two commercially viable annual types of drumstick, which were mainly bred / developed for vegetable pod production (Paliwal et al., 2011b).

Table 7.1 Differences between annual and perennial moringa
(Beaulah et al., 2010)

S.No.	Characters	Annual moringa	Perennial moringa
1.	Method of propagation	Sexual (Seed)	Asexual (limb cuttings, air layering)
2.	Time of flowering	5-6 months after sowing	8-9 months after sowing
3.	Plant type	Bushy and dwarf	Wide spreading
4.	Spacing	2.5 × 2.5 m	5-6 × 5-6 m

Table 7.1 *Contd...*

S.No.	Characters	Annual moringa	Perennial moringa
5.	Population /ha	1600	277-400
6.	Number of pods / tree /year	220 pods / tree / year	600-800 pods / tree / year
7.	Average yield of pods / ha	50 tonnes	27-32 tonnes
8.	Pod characters	Long pods, green in colour	Medium size depending upon the type, green to light brown colour
9.	Cropping period	3 years (one main crop and two ratoon crops)	10-15 years and can be maintained for several years under very good management.

Description of Perennial (eco) Types of Moringa in Tamil Nadu

1. **Moolanur Moringa:** In and around Moolanur, Karur, Dharapuram areas of Tamil Nadu, farmers cultivate this perennial moringa predominantly. It is extremely drought hardy. It sheds entire foliage during extreme drought situations and revives when rain is received. The tree is maintained with a single trunk and unpruned, natural canopy is broader and dense. The tree can be maintained upto 15 years without pruning (Ponnuswami, 2012). Trees flower profusely. Its cropping period coincides during February – April, July-September. Flowering and fruitset are absent during October-January (Beaulah et al., 2010). Pods are short (30-40 cm) highly fibrous and shelf life is more compared to other types. Yield of pods / tree is around 200 kg, pod weight is 120 g. Attack of pests and diseases is less, hence, requires no control measures (Beaulah et al., 2010).

2. **Kodikkal Murangai:** This perennial ecotype was identified by farmers. It is grown predominantly in the betel vine gardens of Tiruchirapalli district of Tamil Nadu. Trees are short statured (dwarf), with small leaves. The pods and leaves are very tasty. It bears short pods (20-25 cm) which are thick fleshed.

3. **Puna Murangai:** This is also a perennial type identified by farmers. It is grown in the home gardens of Tirunelveli,

Kattabomman and Kanyakumari districts of Tamil Nadu. It is preferred for its thick pulp and taste (Ponnuswami, 2012).

4. **Pala Murangai:** It is preferred for its thick pulp and taste pods (Ponnuswami, 2012). Flesh is very white like milk.

5. **Kadu Murangai:** It is perennial wild type with uncertain origin from Sri Lanka. Muthuswamy (1954) indicated that it is actually a wild form of *Moringa concanensis* found in the Niligiri forests of Tamil Nadu. It produces small inferior pods (Kader and Shanmugavelu, 1982).

6. **Kattu Murangai:** It is wild form of *Moringa concanensis* found in the forests of Tamil Nadu. It is highly popular in Tamil Nadu. Trees are shrubby, dense and bushy. Plants propagated by cuttings flower six months after planting. It is suitable for ratooning for every 2-3 years. It bears small and inferior pods. Pod length is 25-30 cm, yield is 400-500 / tree /year.

7. **Palamedu Moringa:** This perennial moringa produces pods of the length of 60 cm, weight of 95-100g and yields 100 pods / tree.

8. **Valayapatti Moringa:** Another perennial type cultivated in and around Usilampatti, Andipatti areas is Valayapatti moringa. The trees produce the pods of 65 cm long, with a weight of 120 g, and yield of 1000-1200 pods per tree. However, it does not produce during winter months i.e. October-December.

9. **Chavkacheri Moringa:** This is an ecotype of perennial Jaffna moringa introduced into India from Sri Lanka and is cultivated in Tamil Nadu. It is popular in Jaffna region of Sri Lanka. It resembles the description of Jaffna. It flowers throughout the year. It bears pods as long as 90-120 cm. It bears heavily – 400-600 pods / tree / year

10. **Chem Murangai:** Another ecotype of Jaffna moringa. A perennial type, introduced into India from Sri Lanka. Tree is medium sized. It bears flowers and pods throughout the year. It bears long pods with red tips (Ponnuswami, 2012).

11. **Jaffna (Yazpanam):** This is also from Sri Lanka introduced into India and cultivated commercially in Tirunelveli and Tuticorn districts of Tamil Nadu for seed oil (Ponnuswami, 2012). It's fruits are 60-90 cm long with soft flesh and good taste. This type can yield 400 pods from second year of planting which increases to 600 pods / tree / year from third year onwards.

12. **Jaffna (Plate 7.1):** This is a perennial land race of moringa introduced into India from Sri Lanka and grown commercially in Tirunalveli and Tuticorn districts of Tamil Nadu. It is propagated by cuttings. It is extremely drought resistant and is capable of growing in back yards, cattle sheds and waste lands. The tree survives on waste water from the house, cattle sheds and rains. When heat stress starts, in the month of February-March flowering also starts and the harvesting period extends upto June-July. Trees are dense and produce several branches and twigs. Branches are cut to get a beautiful shaped tree. Bearing starts from second year of planting. It bears once in a year. Pods are nearly one metre long, ranging from 60-90 cm. It yields 400 pods per tree in the second year of planting which increase to 600 pods per tree per year from third year onwards. Pods are soft, tasty, thick and sturdy. However, it is frequently attacked by hairy caterpillar.

Fig. 7.1 CV. Jaffna fruits.

PKM Varieties

The Horticultural College and Research Institute of Tamil Nadu Agricultural University, Periakulam had developed and released two seed propagated annual moringa types from their germplasm assemblage of 85 moringa accessions. The assemblage contains perennial and annual moringa accessions with heavy fruit bearing, cluster-bearing drought tolerance, dwarf stature and pests and diseases resistance. By judicious breeding programmes including introduction of elite mother plants, evaluation, selection and hybridization, the Institute has released two improved annual moringa varieties viz., PKM-1, PKM-2, with in a span of 10 years, for commercial cultivation during 1990s, which revolutionized the drumstick cultivation in the country. These two seed grown annual drumstick cultivars have replaced 60-70% of the perennial moringa area in South India. Spreading due to their adaptability to varied agro-climatic conditions, precocity and high productivity they represent a lion's share of moringa area and production particularly in marginal and small farm holdings in the country (Sadashakti, 1995; Rajangam, et al., 2001). These varieties have pickd up well in many traditional and non-traditional areas worldwide (Kumar and Pugalendhi, 2010).

Table 7.2 Salient features of annual moringa varieties
(*Source:* Kumar and Pugalendhi, 2010).

Particulars	PKM1	PKM2
Breeding method	Pure line selection	Hybrid derivative
Propagation method	By seed	By seed
Plant stature	Medium, dwarf stature	Medium stature
Pod characteristics	Pods 60-70 cm long with 6.3 cm girth weighing 120g	Pods 150 cm long, girth 8.3 cm, weighing 280 g
Yield	Bears 220-250 fruits per tree, the estimated yield is 50-54 t/ha	Bears 220 fruits per tree the estimated yield is 90-95 t/ha
Growing conditions	Suitable for varied soil types (freely drained) in tropical plains	Suitable for varied soil types (freely drained) in tropical plains

PKM-1 (Plate 7.2)

This is a pure line selection from TNAU released in 1989 for commercial cultivation. It was developed mainly for its long pods and heavy yields. PKM-1 has revolutionized the moringa industry in the country for the simple reason: that it is annual, seed propagated, precocious, widely adapted to various agro-climatic conditions and can fit in any crop rotation and cropping system. Its cultivation is lucrative for farmers as it gives huge income to the farmers. Besides India, it has spread to African countries and other tropics of entire globe. It was found suitable for leaf production and seed oil. It was found suitable for growing as an intercrop as well as growing intercrops in its alleys. It can be grown as intercrop in coconut and fruit gardens during their initial period. It can be intercropped with annuals like chillies, onion, groundnut and sorghum in its alleys (Ponnuswami, 2012). It is world's most successful variety of drumstick as it gives largest yield and income at the shortest period.

Fig. 7.2 CV. PKM - 1 fruits.

The trees of PKM-1 grow to a height of 4-6 metres in a year with 6-12 branches. The trees are robust, bushy, with soft wood. Plants of PKM-1 are precocious and come to flowering within 5-6 months of seed sowing and come to harvest in 7-8 months. Trees flower twice in a year. Pods reach edible maturity within 65 days from anthesis. Pods are 65-70 cm long, thick (6-8 cm), weigh 150 g each, green in colour, tasty, better than other varieties, fleshy, non-fibrous, non-bitter, rubbery, hence unbreakable and straight, so easy to pack. The flesh between the seeds is full.

Pods are harvested during March-August; yield is about 52 t/ha, but its yield potential ranges from 50-54 t/ha. Each plant can yield 200-350 pods per year. Seeds are soft even in late harvest, non-bitter unlike perennial types, rich in seed oil content (38 to 42%) hence PKM-1 is promoted at global level for seed oil. PKM-1 gives economic yields for 4-5 years. Generally, three crops are taken after which the trees are headed back for ratooning, 2-3 ratoon crops are taken after 3-4 years of cropping. The field is replanted with fresh stock.

Tree is free from hairy caterpillar and stem gumming and shows less incidence of pests and diseases.

PKM-2 (Plate 7.3)

This is a hybrid moringa, again from Periakulam, Tamil Nadu. It is seed propagated. It is adapted to varied soil types varying from sandy loams to clay loams with good drainage and suitable for growing in different cropping systems. It can be grown as intercrop in coconut and rubber plantation and tropical fruit orchards. It can be raised as pure crop, as intercrop in orchards and also in back yards. It requires more water, hence cannot be grown in dry areas. Three ratoon crops can be taken before replacement.

Trees are medium tall with more lateral branches. Occasionally, the branches break due to heavy weight of pods, so need propping. Trees flower in about 90-100 days and the first harvest of pods between 170 and 180 days after sowing is possible, i.e. 6 months after planting, during September-October.

Fig. 7.3 CV. PKM - 1 fruits.

Pods are very long (125-130 cm) and often touch the ground and get bruised, girth 8.3 cm, weight 280 g with 70% flesh, less seeded, delicious in taste and greenish in colour. Pods are not straight but wavy and thin. Pods are difficult to pack and transport because of their unusual length and wavy nature. Unlike in PKM-1, the space between the seeds in the pods of PKM-2 is constricted and show wilting impression on the very next day of harvesting. Pods have more flesh than seeds. Pods have good cooking quality and on cooking flesh turns soft and delicious with less fibre. Average number of pods per tree is 220 per year. Each plant can yield 300-400 pods / year. PKM-2 is a heavy yielder giving 98 t/ha. It bears for 9-10 months on plains of India.

KM-1 (Kadu mianmalai-1)

This was developed by a farmer of Pudukoti of Tamil Nadu. It is a pure line selection from annual types, propagated by seed. Trees are bushy, medium tall (Joseph, 2007) and come to bearing 6 months after planting. Fruits are short, 20-25 cm long, with 5.5 to 6.0 cm girth, with 65-85 g weight. The variety bears heavily, 400-500 fruits/plant/year. Trees can be ratooned for 2-3 years and can be replanted by fresh seedlings after 3 years. It is annual type.

PAVM

This is also developed by a moringa farmer (Alagarasamy) of Dindigul district of Tamil Nadu. It became instant hit with hundreds of farmers in Dindigul, Erode and Coimbatore districts of Tamil Nadu. It is propaged by cuttings / air-layering. It requires less water. It starts yielding from 5th or 6th months after planting. Pods can be harvested once in a week. It bears crop for 8-9 months continuously in a year. A single tree produces about 150-200 kgs of mature pods from second year onwards. Annually about 20 t of pods can be harvested. When planted along the hedges more yield was recorded than those planted inside the field. Trees exhibited resistance to pests and diseases and responded well to organic production (Prabhu, 2009).

Coimbatore-1

A perennial type widely cultivated in India. The variety is considered superior in production and quality. It offers two harvests per year. Pods are medium in length (45-60 cm). It bears crops for 8-10 years continuously.

Coimbatore-2

This variety has become more popular in Gujarat. Pods are dark green and tasty. Length of the pods is 25-35 cm and bulky.Each plant yields 250-375 pods for 3 or 4 years. One drawback of the variety is late harvest results in the loss of market value.

Dhanraj (S 6/4)

This variety is from Karnataka, developed at Arabhavi. It is annual type and seed propagated. It is one of the dwarf varieties among the different moringa varieties, growing to a height of 2.0 to 2.5 metres. If flowers early in 7-8 months and bears in 9-10 month from sowing. Its pods measure 35-40 cm long. It is a high yielding variety and yields 250-

300 pods / plant / year. Pods are of good cooking quality. This variety ruled Karnataka upto early 1990's until the arrival of "Bhagya" cultivar.

Bhagya (KDM OI)

Karnataka consumers demanded longer pods and prolific variety. Hence, research for the development of a variety with longer pods and high yielding capacity was initiated at Arabhavi. During 2001, a line from open pollinated seedlings of Dhanraj was selected under the name "KDM OI" and tested against Dhanraj along with other lines over a period of four years. This selection proved its superiority over others in fruit characters and yield at experimental station and in farmers' fields at Bagalkot. Then the selection "KDM OI" was released under the name "Bhagya" by University of Horticulture Sciences, Karnataka. Soon its cultivation has spread all over Karnataka and neighboring states (Brunda Kumari 2014).

Bhagya is well adapted to growing in the hot dry regions in Karnataka. Tree grows to a height of 2-4 metres and long lived (12-20 years) and is pest resistant. It comes to flowering after 100-110 days of planting and pods can be harvested 160-180 days of planting. Pods are medium sized measuring 60-70 cm in length. The yield is 350-400 pods per tree per year.

GKVK – 1,2,3

These are selections developed and released by University of Agriculture Sciences, Bangalore. These were selected for their yield potential and performance. Trees are dwarfs reaching a height of 2.0-2.5 metres. They are early bearers, with profuse fruiting, twice that of good normal trees. Each plant yields 120-200 pods per year with highest record of 400 pods per tree per year.

Anupama

This variety was released by Kerala Agricultural University, Trissur (Suresh Babu and George, 2010). It is early, regular with protracted flowering and high yield. Yield per tree is 30.0 kg and per ha 18.7 t of pods. Fruits are green, measuring 55.5 cm in length and 7.5 cm in girth. Cooking quality of pods is excellent. There is no major incidence of pests and diseases.

AD4

This is a pre-released culture of drumstick from Kerala Agricultural University, Trissur (Suresh Babu and George, 2010). It is seed propagated. Like Anupama, this is also early bearing and regular, with protracted flowering. Pod length and girth are 55.8 cm and 7.4 cm respectively. Fruit colour is reddish green, which turns to dark green at harvest. Yield per tree is 28.5 kg and per hectare is 17.81 tonnes. Cooking quality is excellent. It is not affected by any major pest and diseases.

Rohit-1

This variety was developed by a moringa farmer of Maharashtra after seven years of continuous efforts which led to release of this variety finally. First harvest starts 4 to 6 months after planting and commercial yields continued upto 10 years with two crops a year. Pods are dark green in colour with soft and tasty pulp. Pods are medium sized (45-60 cm long), weighing 60-70 g each fruit. A first year plant produces 585-650 pods per year. From one acre of irrigated and well-drained soil, one can obtain 7-12 t yield.

Konkan Ruchira

This variety was evolved and released by Konkan Agricultural University, Dapoli, Maharashtra in 1992. It was a selection from Vasal local moringa. It produces dark green medium-sized, fleshy and tasty pods with excellent cooking quality (Fugro et al., 1996). Trees are bushy and each tree yields 275 pods (30-35 kg) per year.

SARAGAVA

This is a perennial drumstick and very much popular in Gujarat. Pods are dark green, 40-50 cm long, very stout with high pulp content. It was developed by a farmer.

SARAGAVI

This is also perennial moringa and popular in Gujarat and highly preferred by consumers. Its pods are light green and 40-60 cm long. It was also developed by a farmer.

CHAPTER 8

Seed Propagation

Moringa (*Moringa oleifera* Lamk) can be propagated by either seed (sexual method) of by stem cuttings (asexual / vegetative method). Traditionally perennial types of moringa are propagated by stem cuttings (limb cuttings) while annual drumsticks are propagated by seed. However, seed propagation of drumstick (*M.oleifera*) is extremely difficult as the seeds are over utilized for diverse products including oil for machine lubrication, cosmetics, biodiesel and water purification. Current vegetative propagation of drumstick uses hard wood or softwood cuttings which further threatened the proliferation by seed (Antwi Boasia Ko and Enninful, 2011). Seed propagation has some advantages as well as disadvantages as compared to vegetative propagation.

Advantages

1. It is an easy, cheap and rapid method.
2. It is advantageous in semi-arid and arid areas and also in areas where the depth of water table is a potential growth limiting factor since the plants from seed produce longer tap root and deeper root system than those from cuttings, which explore deep layers of soil and tap water and nutrients for the survival and growth of the plant (Jahn et al., 1986).
3. Seed propagation is possible throughout the year as germination has no restrictive stumbling block other than loss of viability, frost, freezing and water logging environment.

Disadvantages

1. Moringa plants raised from seed produce variation in growth, fruiting and performance. There are reports / complaints from drumstick growers that there would be 15-25% variation in the seedlings of particular variety (Pugalendhi et al., 2010).

2. Since moringa is cross pollinated and heterozygous species, genetic purity cannot be ensured and for maintaining genetic purity, rouging becomes essential which is a costly measure.

3. The plants raised from seeds produce inferior quality fruits particularly in perennial moringa (Ramachandran et al., 1980).

4. Seed derived plants are usually slower to establish inspite of strong root system.

Seed propagation is adopted not only for pod production but also for fodder production in African countries like Ethiopia etc. In Sudan moringa plantations are traditionally raised from seed (Palada, 1996).

Collection (Harvesting) and Selection of Seed

Farmers are advised to procure seed either from his own plantation or from other farmers or from commercial seed producer. In case the farmer intends to gather seed from his plantation he has to follow the following:

1. Identity an elite mother tree(s) showing high productivity and pods of good quality characteristics.

2. Leave some good pods on trees in September for seed purpose

3. Allow them to mature and dry on the tree itself (Plate 8.1).

4. Collect (harvest) the pods at appropriate, time and extract the seeds from completely dried fruits showing brown/straw colour and hairline cracks, before they burst open and spill over the seeds to the floor.

5. Tree dropped seeds are not advisable for seed purpose. However, Kashyp et al., (2009) opined that dropped seeds can also be used for seed purpose. In Ethiopia farmers prefer tree dropped seeds for sowing. Further some farmers use seedlings germinated from the dropped fruits under the tree for transplanting in the main field (Tenaye et al., 2009).

Determination of optimum time for seed collection is an important concern. Delayed harvesting not only induces aging, but also leads to deterioration of seed, reduces the quality of seeds, maximizes the problems due to environmental factors such as high temperature, high humidity, rains, over drying, splitting, attack of diseases and insect pests, birds and animals. Over drying leads to splitting of pods and release of seeds which may be blown off by winds (Copeland and McDonald, 1995). Optimum time, however, depends upon physiological maturity of the pod / seeds. Physiological maturity is the right guide for seed harvesting. Right physiological maturity can be judged by the appearance or change of colour from green to brown/straw colour and hair line cracks in dried pods as well as by the change of colour of seeds (black or brown). Such seeds not only show higher germination, but also store well and longer (Sivasubramanian and Thiagarajan 1997). Harvesting a seed crop at the correct stage not only aims at obtaining seeds of better viability and vigorous potential but also to eliminate the field damage to the seeds by way of over exposure of the mature seeds to the inclement weather conditions as well as to the field pests and pathogens. The chances for amenability to rains and pests and diseases by way of exposure in the field are more in annual moringa as septicidal bunching of the fruit occurs, when the pods are over dried. Thus, the right stage of physiological maturity of pods / seeds assumes greater significance in seed collection in seed production / propagation.

Optimum date for harvesting seed pods is reported to be 100 days after anthesis of flower, when seeds show maximum percentage (98%) of germination. Seeds may be collected during summer months (May to June) depending on flowering date and fruiting as the weather during these months is conducive for drying of pods. But, Kashyap   et al., (2009) suggested April-May months also for collection of seeds.

Individual pods should be selected on the trees based on the length and girth of the pods, each containing 20-25 seeds for seed purpose. Harvested pods may be shade dried for a couple of days and then the seeds are extracted.

Seeds are extracted from the dried pods manually by either split opening or twisting the pods to spill over the seeds. Collect / select only the seeds from the middle portion of the pod, as the seeds at both ends of pod, particularly at distal (tip) end are usually smaller, shriveled and damaged and exhibit poor viability and germination potential. On the other hand, seeds from middle part of the pod show greater physiological maturity and there by higher vigour and germination potential and also vigorous growth of resultant seedlings. If necessary, extracted seed may be re-dried in shade to optimum level of (8-10%) moisture content, particularly when intended for storage.

Then the separated seeds are graded either by specific gravity or based on the seed colour. Grading is carried out with the specific gravity separator. The fraction from 2 and 3 gives higher seedling emergence and vigour. A correlation between seed colour and its quality in respect of viability and germination potential exists. Hence colour grading is advised (Sivasubramanian and Thaigarajan, 1997). Black and brown seeds were found to be superior with higher germination and vigour, than white seeds which was attributed to their optimum physiological maturity. Black seeds recorded 6.7, 51.5 and 24.9% higher germination than brown, white and ungraded seeds respectively (Sivasubramanian and Thiagarajan, 1997). Hence, it is advised to select and use black coloured seeds followed by brown coloured seeds for seed purpose. A good seed should be clean, viable and disease free.

Seed Storage

Since the time of seed collection (April-June) and time of seed sowing (June to August) donot coincide, it becomes necessary to store the seeds till they are put to use for sowing. Under natural conditions (with out storage and seed treatments) moringa seeds can be kept for upto 3 months without much loss in viability and germination. However, to avoid loss in viability and germination potential, seeds must be stored for extended periods. The storability of extracted seeds depends upon the variety, its stage of maturity as indicated by the colour of the seed, moisture content of the seeds, type of container, seed treatment given before storage, storage conditions and period of storage.

Seeds of annual moringa can be stored upto 12 months when freshly harvested / extracted seeds are dried to 8% moisture content and treated with captan @ 2g/kg seed and packed in 700 gauge polyethylene bags (Palanisamy et al., 1995).When the moisture content of the seeds is more than 8-10%, the seed should be treated with a fungicide to avoid diseases and deterioration of viability during storage.

The colour of the seed indicates the level of its physiological maturity. Attainment of black or brown colour indicates full maturity. This was confirmed by the highest expression of viability (93%) by these seeds over 12 months period in black coloured seeds compared to brown (87%) and white seeds (42%). This category seed also possesses highest seedling attributes viz., shoot length, root length, dry matter per seedling and vigour index. Increased maturity leads to better storage of seeds due to higher dry matter accumulation. It was noted that black seeds are more mature followed by brown seeds. Seed vigour, in general decreases gradually during storage but vigour of black seeds was considerably higher than that of brown, white and ungraded seeds in storage (Pugalendhi et al., 2010). The black and brown seeds treated with carbendazim (2g/kg) and stored in 700 guage polyethylene bags maintained 84% germination upto 12 months of storage. Black seeds stored well for 12 months with very little loss in viability in storage followed by brown and white seeds. This is because black and brown seeds are more mature, hence stored well and longer. Irrespective of colour categories, black seeds stored superior by registering higher value for seed vigour followed by brown seeds, while white seeds recorded very poor germination after storage (Sivasubramanian and Thiagarajan, 1997).

As stated, storability of seeds depends on their moisture content. The seed's moisture content should be about 8%. However, the seed moisture content was influenced by the containers used for storage. The seed moisture content for seeds stored in polyethylene bags was 6.7% compared to 8% in cloth bags at the end of 12 months storage period. Increase in moisture content in cloth bags can lead to faster depletion of storage reserves in seeds leading to poor vigour.

Storability of moringa seeds also depends on seed treatment provided prior to storage. Fungicides are usually used as seed treatment chemicals to check/minimize the seed borne disease if any. Poor storage of seeds affects the crop stand and results in reduced yield per unit area. Among the different fungicide seed treatments, seeds treated with captan (2 g /kg seed) could be able to maintain higher seed viability even after 12 months with 60% germination. Higher viability of captan treated seeds and stored in polyethylene bags of 700 guage was due to the fact that the stored seed moisture content was controlled by relative humidity (RH) and temperature, which were checked by polyethylene bags (Palanisamy et al., 1995).

Containers used for storage of seeds also affect the storage of seeds. Germination percentage of seed was higher in polyethylene bags than cloth bags, regardless of seed colour (Sivasubramanian and Thiagarajan, 1997). The maintenance of viability of stored seed in moisture proof containers may obviously indicate through the low moisture build up and consequent decrease in the metabolic activity of seeds (Palanisamy et al., 1995). The increase in moisture content in cloth bags can lead to faster depletion of storage reserves and to poor vigour. Irrespective of the colour categories moringa seeds were stored better in 700 guage polyethylene bags than in cloth bags (Sivasubramanian and Thiagarajan, 1997). Seeds stored in polyethylene bags recorded higher seed quality parameters than those stored in cloth bags. The seeds stored in polyethylene bags showed more than 70% germination upto eleven months, while seeds stored in cloth bags showed germination upto 9 months (Narendra, 2007). Polyethylene bags are better than cloth bags as the seeds in cloth bags absorb moisture and attract diseases resulting in rapid loss of viability of seeds. However, seeds stored in plastic bottles at the ambient conditions lost their viability after 12 months. On the other hand, under cold storage for 24 months the seeds showed reduced viability and quality (Bezerra et al., 2004).

Seed age in storage also influences the germination and vigour of the seedlings. The percentage of germination, rate of germination, plant height and number of branches declined as the period of seed storage extended from 365 days to 440 days (Vijaya Kumar et al., 1999). The fall in the germination percentage and rate of germination is said to be due to the irreversible physical and biochemical changes that take place in the seed due to the initiation of senescence process, attributed to increase in seed age (Justice and Buss, 1978). It was also linked to desorption of moisture from the surrounding atmosphere in open storage (Moravec et al., 2008).

Based on the above, it can be concluded that drumstick seed could be stored in polyethylene bags of 700 gauge for 12 months without much loss in viability of seeds. Kumar and Pugalendhi (2010) however, cautioned that seed should not be stored over longer periods as they lose viability after about a year gradually with seed age and increase in storage period (Caceres et al., 1991).

Seed Treatment

Loss of viability, poor germination and mortality of seedlings in nurseries (both in containers and beds) and late establishment in the field after, planting are the problems encountered in seed propagation of moringa. These problems are attributed to lower moisture content (less than 8%) in seeds, seed and soil borne diseases such as damping off, and root rot (see diseases), pests like termites.It was suggested that these problems can be solved / avoided by soil and seed sterilization with sodium hypochlorite, etc., prior to sowing of seeds. Such practices can ensure high rate of germination, and good seedling emergence and control of seed and soil borne diseases.

Solarization of nursery soil may be adopted for sterilizing the soil. Alternately, soil may be treated with formalin etc. For sterilization of seeds they may be soaked in sodium hypochlorite solution for   10 minutes (Riad et al., 2014).

Fungus accelerates the loss of viability in seeds that have already started deterioration and this can be prevented from further damage by appropriate seed treatment with fungicides like captan and carbendazim. Decrease of viability of seeds during storage was found to be less in seed treated with captan (2 g/kg seed) and stored in polyethylene bags (Palanisamy et al., 1995). Seeds treated with carbendazim (2 g/ kg) and stored in 700 guage polyethylene bags maintained more than 84% germination besides reducing seed borne diseases (Sivasubramanian and Thiagarajan, 1997).

Azospirillum (100 g /500 g seed) helped in maintaining the viability treated seeds compared to untreated seeds, increased seedling height and vigour and showed superiority in higher seedling characters compared to untreated control seeds (SivaSubramanian and Thaigarajan, 1997).

It is best to treat the seeds with a systemic fungicide prior to sowing to prevent damping off seedlings (see diseases).

Seed Rate

Seed rate per unit area (acre / hectare) depends on the variety, mode of raising moringa plantation, spacing adopted in the field, type of soil, location etc. Scanning of available information on the subject indicates variation in the recommendations of seed rate, probably based on the above factors. In most of the cases seed rates were prescribed without reference to mode of raising plantation, type of soil, location etc.

For annual drumstick, the prescribed seed rate is 250 g /ac or 625 g/ha (Kathiresan et al., 1999; Kashyap et al., 2009; Singh et al., 2014).

According to Radovich(2009) about 250-300 g seed will be sufficient to raise plantable seedlings in one hectare. Kumar and Pugalendhi,(2010) suggested a seed rate of 500 g /ha for pod production of moringa.

The seed rate for intensive cultivation for leaf production is higher than that for pod production because of very close spacing in case of former.

Time of Sowing

In tune with seed rate, there is no agreement with regard to the time of sowing of seeds in case of *in situ* planting (direct sowing) or in nursery raising. However, the best time of sowing either in the nursery or *in situ* (direct sowing) is the onset of monsoon rains i.e. June to August it was suggested. The time of sowing mainly depends on location of moringa plantation with prevailing agro-climatic conditions.

Most of the recommendations on time of sowing pertain to South India, more particularly to Tamil Nadu. The lack of suggestions on time of sowing of seeds from other states even in South India, may be due to the fact that no investigations were conducted in other states regarding the time of sowing and may be due to the suitability of Tamil Nadu recommendations and their following in other states.

In South India, drumstick crop is raised in two seasons viz., March-April and July-August depending on the local weather conditions. However, the performance of the crop was not similar under both seasons. Vijaya Kumar et al., (2002), studied the influence of different months of sowing seeds of annual moringa (PKM-1) under the agro-climatic conditions of Coimbatore, Tamil Nadu and concluded that January sowings were more ideal based on the growth of resultant plants. Leaf weight and chlorophyll content were the highest in crops sown in January month with maximum light interception (Vijay Kumar et al., 2002). However, they have not studied the influence of different months of sowing on flowering, fruiting, fruit quality etc. Therefore, it is necessary to study the influence of differing sowing dates (months) on the above in different moringa growing states, to alter the time of sowings to meet the demand of vegetable (mainly pods) in particular month(s) of the year.

The general agreement among the scientists and farmers is to sow/ plant moringa during monsoon season. However, it was cautioned that whatever may be the date of sowing, flowering phase should not clash / coincide with ensuing monsoon rains, lest there would be heavy shedding of flowers and fruitlets and there will be loss of production.

If sown late in November-December, the canopy growth will be less and the tree becomes small statured and start flowering by March, when temperature rises. But the crop canopy becomes weak to bear the abundant pod load. If sown early i.e. by January, the plant becomes too large by March and tend to become vegetative leading to less flower production. Hence the ideal month for sowing of PKM-1 moringa is September in Tamil Nadu (Vadivelu 2010).

The best season for sowing the seed directly in the field is September under South Indian conditions. The time of sowing has to be strictly adhered to because the flowering time should not coincide with the monsoon rains, which results in heavy flower and fruitlet shedding.

Germination of Seeds

Due to genetic variation of the crop, germination can occur within a week or as long as 2-3 weeks, depending on its freshness, maturity, time lapsed after extraction, season (temperature) time of sowing, moisture content in the seeds as well as in the media and shade.

Naturally dropped seeds germinate within 15 days in North India (Tewari et al., 2003). In general moringa seeds take 7-10 days for germination under favourable conditions and take more time when they experience unfavourable conditions (Sharma and Raina, 1982).

Under natural conditions, seed remains viable upto a month, which gets decreased as sowing is delayed. Freshly extracted seeds give 90-95% germination. Germination would be good when they are sown even 90 days after extraction from dried pods (Reddy et al., 2010). Seeds of *Moringa* spp are planted under natural conditions approximately 3 months after harvest in Ghana with acceptable germination (Anon, 2002).

Seeds kept at low temperature do not germinate, whereas germination is faster in warm weather than in cold weather and take about a week to germinate. Under well watered and prepared soil in rainy season (July-August) seeds require 25⁰-35⁰C for germination.

Moringa seeds require shade for germination. Studies by Dania et al., (2014), in Sudan demonstrated that shade had significant effect on germination and seedlings growth variables in drumstick. Medium (50%) shade of one green net layer resulted in faster rate of germination and higher percentage of germination. It also produced erect and strong shoot than extremes of shade indicating the importance of medium shade in the germination and seedlings growth of moringa and that shade should be considered in the nursery conditions.

Improving Seed Germination

Moringa cultivation through seed is problematic because the seeds usually start losing viability within 2-3 months of extraction / storage and thus the germination of stored moringa seeds is a problem and is of great concern to the farmers. This may be overcome by employing seed priming techniques (Nouman et al., 2012b).

Sharma and Raina (1982) reported 60%, 48% and 7.5% germination of moringa seeds sown after 1,2 and 3 months of seed collection. Old moringa seeds have germination problems due to their high oil content and insect attacks (Nouman et al., 2012a).

Seed priming has some benefits: seed priming treatments reduce emergence time and increase final emergence percentage and emergence index (Farooq et al., 2005, 2006, 2008). These also synchronize seedling emergence, which results in uniform stand and improved yield (Du and Tuong, 2002; Harris et al., 2002). Seed priming improves germination and stand establishment and induces tolerance against adverse conditions like abiotic stress especially during emergence and early seedling growth (TedonKeng et al., 2004).

Hydro-priming, matri priming and hormonal priming are effective approaches for accelerating and synchronizing seed germination, emergence and improving plant vigour (Hurly et al., 1991, Wu et al., 1999; Farooq et al., 2006). Moringa leaf extract is rich in cytokinins and its application as a seed proming tool was recently exploited on a range of grasses (Nouman et al., 2012a). Out of the three approaches hydro-priming proved more effective than other two approaches. It was reported that 12 hours of hydro priming (soaking in water) increased the germination of moringa seeds, collected from Nicargua and Kenya

(TedonKeng et al., 2004). Pot studies indicated that 8 hours of hydro-priming of moringa seeds effectively improved the emergence, shoot biomass and chlorophyll b contents, while root biomass, chlorophyll a content and mineral contents were effected by moringa leaf extract priming (for 8 hours) and both priming strategies were effective in improving β-carotene content in leaves. These two approaches are organic, inexpensive and environmentally friendly techniques that require little input, but give good results (Nouman et al., 2012b). According to TedonKeng et al., (2004) the germination of fodder moringa seeds can also be increased by hydro-priming.

Moringa can be grown directly or through nursery plants. In order to have quick results it is best to go in for nursery plants. If water is available for irrigation moringa can be seeded directly and grown any time during the year.

Direct Sowing of Seeds in the Field

Direct sowing of seed in the main field is the most common method for *M.oleifera* in India (Kumar and Pugalendhi, 2010) and in Southern Ethiopia for *M.stenopetala* (Tekolay and Demel, 1995). It was considered best to plant the seeds directly where the tree is intended to grow, over transplanting of seedlings, as the young seedlings are fragile and often can not survive transplanting shock. Direct dibbling of seeds assures accelerated and faster growth of plants. Direct seeding is preferred, when plenty of seed is available at cheaper cost, and labour is limited. Direct seeding is possible because the germination rate of *M.oleifera* is high (>90%). Transplanting of seedlings on the other hand, gives flexibility in field planting, but requires extra labour and cost in raising seedlings. Production of seedlings in seed beds / containers is very time consuming; but plants can be better protected from cattle and pests. Transplanting approach is used in areas where soil erosion is a problem.

Area for direct seeding should have light and sandy soil but not clay or water logged (see soil). After thorough preparation of the field (see preparation of field) pits may be dug up at pre-determined spacing. The pit size depends upon soil type. The pits serve to lose the soil and help to retain moisture in the root zone enabling the seedling roots to develop rapidly and deeply. In light soils 30-60 cm^3 pits and in heavy soils 90 cm³ size pits may be dug 20-30 days before dibbling the seeds. Back fill the pits with loose soil, plus compost or manure to help the tree growth, even though moringa tree can grow in a poor soil and without manuring. Filling the pits with fresh top soil is beneficial, as it contains beneficial microbes that can promote more effective root growth. However, moringa trees are very strong and could easily grow in unfertilized soil as long as the ground is soft. Water the pits copiously for settlement of filled soil in the pits and be moist for seed germination or wait until a good rain is received for dibbling seed.

Generally, when fresh seeds are utilized, they do not require any pre-treatment for encouraging germination and seedling emergence. But in some cases hydro-priming may be needed. Soak the seeds in water for 24 hours for imbibing water by seed it needs for germination. After this treatment, the seed is mixed with *Azospirellum* @ 100 g/ 500 g of seeds. Seeds may also be treated with rice gruel biofertilizer slurry, shade dried for 30 minutes and then sown in the pits (Beaulah et al., 2010).

Dibble 3-5 seeds in each hole at a distance of 2-3 cm apart, not deeper than 3 times the width of the seed (approximately 1.5 cm depth). When the seeds are expensive or difficult to procure dibble 1-2 seeds per pit (Kumar and Pugalendhi, 2010).

Best time for sowing seed directly in the field is September under South Indian conditions but never plant seeds when evening temperature goes below 13⁰-14⁰C.

When the seeds germinate thin the seedlings to one strong and vigorous one and remove the rest without disturbance to either of the seedlings after two weeks of germination or when they reach a height of 20-30 cm. In case neither of the seeds germinate, the pit may be opened to check if there is a localized insect attack (termites / root grub) in which case the pit soil is replaced with fresh soil or treated with neem oil or neem leaf solution mixed with soapy water and seeds are resown (Armella –de-Saint and Broin, 2018). The extra seedlings if suitable may be used for gap filling if any in the field. However, safest method would be to raise 50-60 seedlings extra in suitable containers for gap filling for uniform (age) stand.

After sowing keep the soil moist, so that soil will not dry and check/ choke the emergence of seedlings, but it should not be too wet, lest the seeds / roots rot. Similarly after the emergence of the seedlings water the pits to keep the soil moist so the emerged seedlings grow fast and vigorously.

Raising Seedlings in Nursery Beds

Select the site for raising seedlings (in nursery beds) under shade to prevent exposure to excessive sunlight and near a water source. The soil of the nursery beds should be light, friable and well drained. Before the onset of rainy season, nursery site should be prepared well by ploughing, harrowing and leveling and finally made into beds. Alternately, first beds may be made, and prepared well by digging, removal of stones, pebbels and plant debris etc. The nursery soil is subjected to solarization or treated with formalin to kill all the harmful microbes in the soil. The beds should be of one to two metre width and convenient length as rised beds (15-20 cm) particularly in heavy soils against excessive rains and water stagnation (Pande et al., 2013). If seedlings are started in rised beds, the soil should be sterilized by burning 3-5 cm layers of paddy straw or other organic matter on the bed. The burned ash adds minor amounts of nutrients such as phosphorus and potassium.

Preferably sow freshly extracted seed for rapid and higher germination from July to October, not to coincide with rainy season, which will affect seed germination (Beaulah et al., 2010). Sow 2 or 3 seeds at a distance of 10-25 cm between rows and 2-3 cm between the seeds in lines (furrows) at a depth of 1 to 1.5 cm. After sowing the seeds, the beds are covered with a mixture of well rotten compost plus friable soil plus sand (2:1:1) and mulched with dry paddy straw / dried leaves to protect seedlings from pests, heavy rains and harsh sunlight. This mulch may be removed as and when seeds germinate and seedlings emerge from the ground.

The nursery beds may be irrigated before sowing. The beds are watered every day after sowing till germination and there after once in 2-3 days depending on the need till they are lifted for out planting in the main field.

Raising Seedlings in Containers

When it is not possible to plant seeds directly in the main field and not possible to raise seedlings in the nursery beds, seedlings are raised in containers. Seedlings can be raised in containers like individual trays, individual pots, or plastic bags. Use of divided trays and individual containers is preferred because there is less damage to the seedlings when they are lifted and transplanted. Container grown seedlings are easy to transport. There is no elaborate preparation of soil for raising the seedlings. Seed rate is also considerably less. The most suitable containers are polyethylene sacks (bags), because of their light weight, transparency and moisture holding capacity. Seedling growth is rapid in polyethylene bags. Seedlings reach a height of 20-30 cm in six weeks after sowing in containers and become plantable (30-50 cm height) in eight weeks after sowing (Beaulah, et al., 2010).

Size of polyethylene bags varies from place to place and recommendation to recommendation as a result there is no uniformity. Sizes may vary from 5" x 9" (Sreenivas Rao and Hariprasad Rao, 1996) to 18" x 12" cm. Before filling the bags, drainage holes may be made at the lower side of the bag. Again the components of pot mixture vary widely. It was suggested to fill the bags with red soil plus FYM+sand (2:1:1) (Sreenivas Rao and Hariprasad Rao, 1996) or light soil mixture i.e. 3 parts of soil to one part of sand (Anon, 2012a) or a mixture of 67% peat moss and 33% coarse vermiculate of AVRDC. Arrange the bags in a slightly shaded area, where they are protected from heavy rains and glaring sun. Also arrange the filled bags in rows for easy watering and weeding etc.

The best suitable season for sowing seeds in containers was suggested as August-September in Tamil Nadu (Pugalendhi et al., 2010). Seeds may be sown during April for planting in the rainy season (Kathiresan et al., 1999). The time of sowing should be strictly adhered to avoid coincidence with monsoon rains and resultant shedding of flowers and fruit lets.

Before sowing the seeds may be soaked in water overnight, cracked and kernels without shell are sown for early, faster and better germination of seeds. The seeds (kernals) are treated with *Azospirellum* culture (100g/500 g seed) for early germination, increased seedling vigour and yield (Gomathi et al. 1999).

Sow the fresh seed within 2-3 days of extraction @ 2-3 seeds per bag at a depth of 0.5 cm. Keep the soil in bags moist, but not wet. Irrigate with rose can or hose pipe daily till germination. Thereafter once in 2-3 days depending upon the dampness of the soil in the bags or as and when required. Watering should be done with extreme care to avoid collapse of fragile seedlings due to pressure of watering. Collapsed seedlings if any must be supported with stakes.

Germination occurs within 7-12 days after sowing. Young seedlings must be nursed for 4-6 weeks before planting. Seven to ten days after germination, thin them to one strong and vigorous seedling.

Though seedlings would be ready for transplanting after 30 days when they reach a height of 30 cm seedlings may be kept in polyethylene bags for want of favourable situation for transplanting, upto 60 days (Kathiresan et al., 1999). Polyethylene bags may be removed by cutting longitudinally on the side, and seedling is removed without damaging the root system of seedlings.

AVRDC suggests to use 50-cell trays with 3-4 cm width and depth. The trays are filled with potting mixture that has good water holding capacity and good drainage. Peat moss, or commercial potting soil mix or a potting mixture prepared from soil, compost or rice hulls and vermiculite or sand may be used. AVRDC uses a mixture of 67% peat and 33% coarse vermiculite. When non-sterile components are used it is sterilized by auto-claving or backing at 150^0C for 2 hours.

Seedlings in trays may be grown under or in a screen house with 50% shade. Two to three seeds may be dibbled in each cell of the tray. One week after germination, thin to the strongest and healthy seedlings.

Seedlings are to be irrigated every morning or as needed talking care to keep soil moist but not wet, using a fine mist sprinkler to avoid soil splash and plant damage.

Moringa seedlings raised in containers will be ready for transplanting after a month of germination when they reach a height of 50 cm.

Transplanting the seedlings in the rainfed culture should be done in June-July when the rain fall pattern has stabilized. If the irrigation facilities are available, transplanting can be done earlier. Transplanting should be done in the morning or late in the evening to avoid dehydration of seedlings and stress.

Seed Production in Drumstick

Since the introduction of annual moringa area under drumstick is increasing year after year in India. Since, annual types (PKM-1 and PKM-2) are propagated by seed, there is a great demand for seed to raise new plantations and area expansion under moringa. There is, however, dearth of quality seed, supply of seed is not keeping pace with the demand. It is estimated that the annual requirement of moringa seed for further expansion of area and to replace the existing plantation after 3-4 years of cropping, is over 600 tonnes per year. This does not include seed demand for other purpose. Farmers from different states were procuring drumstick seeds at huge cost (Rs.2500-3500 per kg or more) for several years in the past from Tamil Nadu Agricultural University (TNAU), Coimbatore and Periakulam. However, TNAU could not meet the ever increasing and huge demand for seed. Hence, TNAU decided to take the help of private farmers to produce seed on their farms (on contract basis) and supply seed to TNAU. According to the policy of government of India of "seed village" the farmer is required to produce his seed himself on his farm itself and be self-dependent. In tune with the above policy, TNAU has developed and standardized the seed production technology in drumstick and released the package of practices for commercial seed production to farmers. It is essentially similar to commercial production of pods, with slight changes. By adopting the following the seed production technology, seed producers can embark upon large scale production of seeds on scientific basis to get seeds of genetically pure, vigorous and germinability. The salient features of seed production technology of moringa, as standardized by TNAU are as follows:

Seed production is a specialized job and needs more understanding and skill and attention to maintain the genetic purity of a variety. Hence, the seed production strategies for annual drumstick are important to produce high quality seeds with high genetic purity and high standards, to ensure uniform crop with higher yields.

Climate and Soil

Climate and soil requirement is similar to that of production for pod production. Moringa comes up well in a wide range of soils. A deep sandy loam soil with a pH of 6.5-8.0 is optimum. Highest seedling vigour can be obtained in red soil compared to black loam, loamy or sandy soil. Drumstick is a tropical plant and grows well in the plains. However, it also grows in subtropical climate. It is predominantly a crop of dry and arid tracts. The optimum temperature for better growth and production is 25^0 to 35^0C. However, it is susceptible to frost and high temperature exceeding 40^0C, which cause flower shedding, (Ponnuswami, 2012). Therefore, such soils should be avoided.

Digging and Filling Pits

In a well prepared field, pits of 45 cm³ or 60 cm³, depending upon the type of soil are dug out at a spacing of 2.0 × 3.0 meters or 3 × 3 m on either side. The pits are filled with 15 kg FYM/compost per pit plus NPK at 100:200: 50g per plant as basal dress.

Seeds and Sowing

Genetically pure genuine seeds must be used for seed production, procured from Government agencies / institutions, to start the commercial production of seeds. Later seeds from own farms can be used. A seed rate of 500-600 g per hectare is required. Fresh seeds are sown during July to September under South Indian conditions particularly in Tamil Nadu.

Before sowing, moringa seeds may be treated with *Azospirellum* culture @ 100 g per 500 g of seed for early germination and increased seedling vigour, growth and yield. Then the seeds are sown in pits @ 2-3 seeds per pit at a depth of 2.5 to 3.0 cm, which germinate within 7 to 12 days after sowing.

Alternately, container grown / nursery raised seedlings may be utilized for seed production. Such seedlings will be ready for planting one month after sowing. An additional number of 75-100 plants may be raised in poly bags for gap filling purpose (Ponnuswami, 2012).

Isolation Distance

Drumstick is a cross pollinated crop. Hence, to achieve genetic purity of seed of a selected variety, the plants should be grown in solid blocks with an isolation distance of 500-600 metres at least between the varieties to avoid cross pollination and contamination of seeds, for commercial seed production. It is advisable to grow preferably only one variety as a monoblock garden allowing free insect pollination. For effective pollination and good fruit set and yield set up 2-3 bee hives per hectare in the field.

After Cultivation

Gap filling may be done within a month to maintain the population of plants. Pinching of the main shoot when the plants reach a height of 75-150 cm to facilitate more branching, reduce the plant height and to get higher yield should be done between 60 and 90 days after sowing / planting (Vijay Kumar, 2000). Ponnuswami (2012) advised two pinchings at 20-25 days interval.

Manual or mechanical weeding may be done to keep the field weed free. Alternately intercrops like tomato can be grown to reduce the weed growth besides getting additional income in the early period, Mounds are to be formed around the tree trunk upto a height of 30-40 cm from the ground level to avoid breakage of branches due to crop load and high wind.

Irrigation

Moringa generally does not need watering except during hot weather, when they may be irrigated once in a week for three months after planting and once in ten days thereafter. Water stagnation should be avoided, there will be flower drop when the soil is too dry or too wet. Hence optimum moisture should be managed. Yields can be doubled by drip irrigation compared to rainfed crop. Drip irrigation @ 4 litres per day can enhance yield by 57% as compared to rainfed crop (Raja Krishna Murthy et al., 1994). During flowering irrigation should be restricted to avoid flower dropping, while liberal irrigation should be given during pod development.

Nutrition Management

Adequate nutrient management is essential for the production of good quality and higher yield of seeds. According to Sunthanapandian, et al., (1989) moringa requires 44:16:30 g of NPK per tree at the time of pinching (75 days after sowing). Nitrogen @ 44 g / tree must be applied as top dressing at first flowering (150-160 days after sowing). Palanisamy et al., (1995) suggested basal application of 15 kg FYM per pit and top dressing of NPK @ 100 : 200 : 50 g / tree in 3 splits at flower initiation, flowering and fruit development stages and a foliar application of 3-4 sprays of NAA 20 ppm at flowering at 10 days interval. Of late, Ponnuswami (2012) stated that application of 100g urea, 100 g super phosphate and 50 g muriate of potash per pit should be given three months after sowing, followed by application of 100 g of urea per plant at the time of flowering. He further advised the application of NAA @ 20 ppm followed by 2 kg urea plus 4 kg super phosphate plus 400 g micro nutrient mixture sprays per hectare, 3-4 times at 10 days interval and opined that with this measure 3 ratoon crops can be taken up without reduction in quality of seeds.

Pest and Disease Management

The major pests of drumstick are fruit fly, aphids and Jassids which can be controlled by carbaryl spray (0.2%) or any other systemic insecticides. Out of the above, pod fly (*Gitona distigma*) is an important and destructive pest causing damage and direct loss in seed crop. Pod fly can be effectively managed by (i) application of Fenthion 80 EC (0.04%) during vegetative and flowering stages, (ii) application of nimbicidine (0.03%) at 15 ppm during 50% fruitset and 35 days later, (iii) soil application of neem seed kernel extract (NSKE) at 2 litres / tree at 50% fruitset and, (iv) weekly removal of infested fruits (Anjaneyamurthy and Regupathi, 1992; Ragumoorthy and Subba Rao, 1998) (see insect pests). The disease like root rot can be controlled by root drenching with copper oxy chloride (0.2%) (see disease).

Rouging

Being a cross pollinated and heterozygous crop, there are chances for variation even with in a variety / clone. So the plants should be watched closely and thoroughly for characters on the selected variety and variants or off types should be rouged at different stages like, vegetative growth, pre-flowering, flowering and fruiting stages. Based on the plant stem characters, during early stage, the rougues (off types) should be completely up rooted and the resultant gaps should be filled promptly to maintain the stand. During pod development and maturity stages based on pod characters the process of rouging should be continued, for example, pods showing pink pigmentation, instead of green colour and the pods less than 70 cm long and tribased shape should be rejected and only those pods more than 70 cm long and cylindrical in shape and green should be retained and harvested in case of PKM-1. Similarly when seedling bears entirely off type pods considering a variant, it should be rouged out. Similarly, when seedlings reach 20 cm height remove any weaker ones and fill the gaps.

Maturity and Harvesting (Plate 8.1)

Seed propagated annual moringa flowers in about 100-110 days and the first harvest of fruits can be picked between 160-180 days after sowing. Seeds of annual moringa attain physiological maturity at 70 days after anthesis (DAA) with maximum germination and vigour. They attain harvestable maturity 100 DAA when the fruits start drying and develop hairline cracks. Delay in the harvesting leads to shattering of pods and loss in seed quality. For seed purpose mature black or brown coloured and tree dried pods should be harvested. At this stage the seeds have attained full physiological maturity and possess high viability and germination potential (upto 98%).

Fig. 8.1 Drumstick pods left on a tree for seed purpose.

Fruits from February – March flowering, maturing in summer months are suitable for seed purpose, when it is easy to dry and the incidence of pests and diseases is comparatively less. Further, such facilities are not available in other season. According to Palanisamy et al., (1985), the optimum age for seed harvesting is 100 days after anthesis (DAA) when the highest germination (>90%) occurs. A delay in harvest date beyond 100 DAA results in seed deterioration. In view of the above, harvesting of black or brown coloured seeds is preferable as they lead to recovery of good quality seed with high germination potential.

Seed Extraction (Plate 8.2)

Tree dried pods showing hair line cracks should be harvested before they burst and spill over the seed. The harvested fruits should be shade-dried with good ventilation for a couple of days and then the seeds are extracted manually. The seeds from stalk end (proximal) and distal end should be rejected and those in the middle portion of the fruit should be gathered as they are of high quality with higher potential of viability and germination (>90%). Seeds from stalk end (proximal) and stylar end (distal) of the fruit are often white in colour, smaller shriveled and damaged ones with low viability and germination, hence should be discarded (Plate 8.2).

Fig. 8.2 Extracted drumstick seeds.

Ratooning

Three ratoon crops can be taken for 3 years without reduction in seed quality (Sivasubramanian, 1996). However, as per the certification procedure, one ratooning is allowed for certified seed production. However, three or more ratoonings can be practiced for truthfully labeled seed production (Ponnuswami, 2012). For ratooning the trees are cut at a height of 90-100 cm and the same cultural operations of main crop are to be followed.

Grading of Seed

First small, ill filled and damaged seeds can be separated by using 24"/64" round perforated metal sieves. Then the seeds are graded according to the colour of seed or specific gravity. Colour grading is easy and does not need any equipment. Black coloured seeds (first grade), followed by brown coloured seeds (second grade) are superior with higher seed quality attributes such as germination and vigour than white seeds.

Grading of moringa seeds can be done with specific gravity separator. The fraction from 2 and 3 gives higher seedling emergence and vigour (Ramesh Kumar et al, 2014).

Seed Yield

Seed yield varies between 1000 and 1200 kg per hectare depending on the maintenance of seed production farms. Within 3 years, a young tree produces 400-600 seed pods annually (Palanisamy et al., 1985). According to Ponnuswami (2012) annual moringa yields 200-250 pods per tree per year. Each pod contains 10-13 first grade seeds giving a seed yield of 2000-3250 seeds per tree per year i.e. 600 to 1000 g per tree. Palanisamy et al., (1985) recorded a seed yield of 150 kg per hectare in PKM-1.

Seed Storage

Seed storage is necessary till they are used for sowing and / or disposed off by sale. The seeds can be stored at low temperature and containers after treatment with suitable fungicides. The optimum temperature for seed storage is 20^0-25^0C. Annual moringa seeds can be stored up to 12 months when freshly extracted seed are dried to 8% moisture content and treated with captan @ 2 g /kg seed and packed in 700 guage polyethylene bags (Palanisamy et al., 1995). The black and brown coloured seeds treated with carbendazim (2 g / kg seed) and stored in polyethylene bags stored longer than in cloth bags and maintained more than 84% germination upto 12 months (Sivasubramanian and Thiagarajan, 1997). Higher viability of stored moringa seeds can be achieved with a storage RH of 75% at 10% moisture content in seeds (Sivasubramanian 1996).

Seed Certification Standards

For certification of seeds, only main crop is allowed and for labled seeds main crop with two ratoons are allowed. The minimum germination should be 70% and 1000 metres isolation distance is prescribed for breeder seed and 500 metres for certified seed in moringa (Ramesh Kumar et al., 2014).

Table 8.1 Field Standards seed for certification (Ponnuswami, 2012)

Factor	Maximum field standard	Permitted (%) CS
Off – types	0.1	0.2
Plants affected by seed borne pathogens	0.1	0.5

Note: Standards for off types should be met at and after flowering stage.

Propagation by Limb Cuttings

Although moringa can easily be raised by seeds, some prefer vegetative propagation through limb cuttings for raising new plantations because of the following reasons:

1. Being a cross pollinated and heterozygous plant, seed propagated moringa plants exhibit variation to an extent of 15-20% which is not the case with vegetatively propagated plants. Hence, asexual propagation through stem cuttings is preferred over seeds particularly in case of perennial types to maintain and avail genetic purity and dwarfing of trees (Muthuswamy, 1954; Verma and Chaarasia, 1993).

2. Plants raised from seeds produce interior quality fruits. Therefore, for better quality and uniform fruiting, limb cutting's propagation is preferred (Saroja and Dwivedi, 1993).

3. Seed propagation is threatened by high demand for seed for other purposes (like seed oil), consequently the availability of seeds became scarce and expensive world-wide as the present supply does not meet the demand for multiplication of crops and other uses. This further requires more planting material (Anon, 2010c).

4. In general vegetative propagation is an alternate means to increase plant stock. This is also true in case of moringa. Instead of over reliance on the limited supply of seed for propagation, it is better to adopt vegetative propagation which would offset the problem of seed shortage, increase resources and contribute to more sustainable local use (Antwi – BoasiaKo and Enninful, 2011).

5. The vegetative propagation of perennial types of moringa is necessary to obtain true to type stock (plants) with uniformity in production and quality of produce and also they are maintained for 8-10 years at a stretch in the field unlike annual types, which are removed after 3-4 years from the field (Thamburaj and Natarajan, 2010; Singh, 2011).

6. With the demand for bulk quantities of moringa to increase the number of trees per unit area the farmers started increasing the number of trees through limb cuttings of perennial types in late 1990s after the introduction of "Jaffna" variety (Anon, 2012a).

Vegetative propagation is necessary to obtain true-to-type plants with uniformity in yield and quality. Studies conducted on comparison of different methods of propagation of moringa var. PKM-1 (seed, limb cuttings, half-rooted trees and fully rooted trees) revealed that best propagation method was limb cuttings in terms of genetic purity of the variety, growth, quality and yield (Swati et al., 2013).

However, in spite of all the above, propagation by limb cuttings is beset with certain advantages and disadvantages

Advantages

1. Limb cuttings are easiest to root hence, chances of failure are less (Muthuswamy, 1954).

2. Plants from cuttings grow faster

3. Vegetatively propagated (by cuttings) plants exhibit purity and genetic uniformity in yield and quality of produce, which is the main intension of vegetative propagation (Doerr and Williams 2007-09).

4. They come to bearing earlier, particularly in case of perennial type of moringa, in comparison with seed propagated perennial plants. The plants from cuttings will commence to bear in the 5th or 6th month of planting (Muthuswamy, 1954).

Disadvantages

1. Plants from cuttings lack tap root and have a shallow root system, which makes the trees more susceptible to drought and winds (Doerr and Williams 2007-09).

2. They develop side branches which are prone to breakage due to heavy winds, because of brittleness of limbs (Singh, 2011).

3. Plants propagated from felled shoots, show higher mortality of planting material in spite of development of roots and more branches (Antwi – Boasia Ko and Enninful, 2011).

4. Further, plants from cuttings are sensitive to termite attack (Kumar and Pugalendhi, 2010).

5. Vegetative propagation of *Moringa oleifera* is possible only with large cuttings which greatly limits the use of this technique for medium or large scale plantation.

6. It is not suitable for propagation of annual types

Problems

While propagating moringa through cuttings, the following problems are encountered:

1. Limb cuttings often fail to sprout, particularly when cut and planted during improper season, due to damage of delicate buds, located on the very surface of bark of cutting during collection, handling and transportation (Beaulah et al., 2010) because of which stem cuttings fail and die when planted in the field from nurseries.

2. The productivity of the trees from cuttings is more erratic, especially under poor management without fertilization (Beaulah et al., 2010).

3. Though trees raised from cuttings are precocious and bear fruits in 6-8 months after planting, yield is generally low in the initial couple of years, but from third year of planting, yield rises (Ramachandran et al., 1980).

Source of Cuttings

Elite mother trees that are healthy, produce genetically uniform crops with high productivity and quality of produce and free / tolerant to important pests and diseases are the best and ideal sources for collecting / taking cuttings for propagation. Such elite tree should first be identified based on observation and record, as in the case of seed propagation. Limb cuttings from such trees can directly be sourced for collection and propagation. However, excessive collection of limb cuttings from a single tree should be avoided, which may damage the trees, which take a couple of years for recovery and to come to bearing.

Alternately elite trees are pollarded to a height of 90-100 cm from the ground which produces 2-3 shoots and these shoots are utilized for propagation (Seemanthini, 1964; Peter, 1978). Branches (shoots) from pollarded trees grown for pods and fodder are excellent for propagation of moringa. Branches put forth after harvesting of fruits can also be used for propagation (Saroja and Dwivedi, 1993).

Type of Wood

Generally hard wood stem cuttings are suggested for propagation of drumstick (Reddy et al., 2010; Kumar and Pugalendhi, 2010), because they show physiological maturity and sprout best with highest number of sprouts (Antwi-BoasiaKo and Enninful 2011). In a study, however, they achieved success in vegetative propagation in moringa both with hard wood and semi-had wood cuttings treated with NAA. However, it is better to avoid using semi-hard wood and young green tissues for propagation as they wilt and fail to root.

Selection of Cuttings

Limb cuttings are advised for vegetative propagation of drumstick. Straight, mature hard wood branches (shoots) are cut off one metre from the tip of the branch just below the node. Then defoliate the shoots along with tender growing end of the branch. Cuttings for propagation should be from at least one year – old wood (from previous years) (Beaulah et al., 2010).

Size of Cuttings

A perusal of the available information on the size of cuttings revealed that there is no uniformity in the size of the cuttings (length and girth) for propagation. There is variation between different states, countries and authors.

In India, long limb cuttings of 100-150 cm and 14 to 16 cm circumference (girth) are generally used in case of perennial types (Palada, 1996). Earlier, Muthuswamy (1954) suggested the use of limb cuttings measuring 3-4 feet in length and about 6-8 inches in thickness in Tamil Nadu. On the other hand, 90-100 cm long and 5-8 cm thickness cuttings are used in Andhra Pradesh (Rangappa, 1980). The branch cuttings of about 1-2 meters long and 10-20cm girth are used in Dehra Doon (Saroja and Dwivedi 1993). Current vegetative propagation methods, in Ghana use 45 to 150 cm long hard wood or soft wood cuttings for propagation of moringa (Antwi-Boasia Ko and Enninful, 2011).

Hard wood cuttings of 20-30 cm length with one centimetre diameter are also used. Antwi-Boasia Ko and Enninful (2011), however, achieved propagation with 30 cm long hard wood as well as semi-hard wood cuttings in Ghana, with NAA treatment.

Collection of Limb Cuttings

Straight lateral shoots with buds, from elite, healthy mother trees should be collected (Tenaye et al., 2009). Cuttings may be collected during rainy season (August-September), since at this time no fruit will be on the trees and safe as cutting of branches does not cause any loss/damage to the trees (Rangappa, 1980). Ponnuswami (2012) advised to cut off branches after the trees have stopped production of pods. This, according to him, will promote fresh growth and the cut branches provide excellent cuttings for growing new trees.

Preparation of Cuttings

Before planting, the cuttings should be properly prepared by (i) trimming the green wood without damaging the bark of the cutting, (ii) curing the cuttings for three days under shade, iii) treating the cuttings with IBA 500 ppm for higher rooting percentage (Beaulah et al., 2010).

Seed Rate and Time of Planting Cuttings

Sufficient information is not available on seed rate i.e number of cuttings per unit area. The number of cuttings required for planting depends on the type of moringa, spacing to be adopted etc. The following seed rates were however, suggested. For perennial types it is 600 cuttings per hectare (Pugalendhi et al., 2010) while it was 640 cuttings per acre for annual types of drumstick.

Limb cuttings will root without much care, but they grow best when planted at the start of the rainy season and any other time when the weather is mild. One should avoid planting cuttings in very hot or cold weather (Ponnuswami, 2012). Muthuswamy (1954) suggested planting during July-August in Tamil Nadu. Cuttings are planted during the time between June to October in Andhra Pradesh (Rangappa, 1980). Saroja and Dwivedi (1993) and Palada (1996) also suggested start of rainy season as the best time for planting of cuttings in the field. Rainy season planting of cuttings causes early rooting and helps early growth (Beaulah et al., 2010). Planting of cuttings during wet period is vital for rooting and survival of the cuttings (Antwi-Boasia Ko and Enninful, 2011). In North India, rooting success was better during spring months than either in summer or rainy season or in the cooler winter months (Beaulah et al., 2010). In Saudi Arabia, limb-cuttings 1-2 metres long are planted from June to August (Mridha, 2015).

Field Planting of Cuttings

Moringa limb cuttings can be planted directly *insitu* unlike other horticultural crops, which are first made to root in nurseries before they are set out in their final place (Muthuswamy, 1954). Of late however, limb cuttings are allowed to root either in nursery beds or in containers or in screen houses, when there is adequate time between raising rooted cuttings and planting in the field (Ponnuswami, 2012).

In case of direct planting of cuttings, plant them in a well prepared sandy soil in pits. Pits of 50-100 cm³ depending on the type of soil are dug at 5 × 5 m spacing, at the beginning of monsoon rains and refilled with appropriate soil + manure / compost mixture. While planting in pits, burry 1/3rd of the bottom of stem of cutting in the soil (Kumar and Pugalendhi, 2010). After planting the cuttings, press the soil around the base of the cutting to avoid falling, while watering or due to winds. Then form a slight dome or cone-shaped mound sloping down from cutting to prevent water touching the base of the cutting. In India, the custom is to put a handful of cattle dung on top of the cutting to avoid loss of moisture by transpiration from the cut end and to protect the cutting from pests and diseases and sun.

Rooted cuttings from nursery beds or containers are transplanted in the field after 2-3 months of planting in the nursery (Ponnuswami 2012). Rooting of cuttings is slow in nurseries in which case it is advisable to apply phosphorus to soil to encourage the development of roots. Cuttings in nurseries strike roots within 4-6 weeks of planting.

Appearance of sprouting on cuttings indicates that cuttings have rooted and survived. Protect the survived cuttings form goats and cattle, water the cuttings judiciously, but do not over water especially in heavy soils lest roots may rot (Doerr and Willaims, 2007-09). Under moderate temperature watering should be done during the initial days of planting just to optimum levels to avoid root rot (Ramesh Kumar et al., 2014). Add balanced fertilizers or compost to encourage root development and growth of plants. Under good management and favourable conditions, plants from cuttings start bearing pods 6-8 months after planting with less yields during the 1 or 2 initial years, but regular bearing commences from 2nd and 3rd year onwards and continue to bear several years.

Micro Propagation

Traditionally trees of moringa (*Moringa oleifera* Lam) have been multiplied for new plantation by sexual (seed) and asexual (cuttings) methods. However, there are some demerits in both of these methods.

In India, Thailand etc. moringa especially annual types are propagated by seed. This method is slow, takes between 3-14 days for germination and results in one plant per seed. Being a cross-pollinated and heterozygous crop, the progeny exhibits variation in genotypes, hence in their phenotypes resulting in variation in fruit production (pod) quality and nutritional values (Riyathong et al., 2010). Sexual propagation of the *Moringa* spp. is tedious and not possible in some species. It requires large number of individual plants for cross pollination. Flowering does not commence until a critical size/stage is attained. Further seed formation is often unreliable and the seeds have poor viability and germination if not planted shortly after maturity and extraction from pods, although they have no dormancy (Kantharajah and Dodd, 1991). Moreover, seeds are not available in required quantities, as they are utilized for other purpose like extraction of seed oil (see seed propagation).

Perennial moringa types are traditionally propagated vegetatively by limb cuttings and this method is slow, inefficient and large number of limbs (branches / shoots) are required, which may damage and even kill the mother trees (Kantharajah and Dodd, 1991). Rooting and survival of limb cuttings are low. Moreover, availability of planting material (cuttings) is also restricted as they are to be obtained at the cost of production of pods. Further, the plants from cuttings develop side branches which are soft and brittle and hence are more prone to breakage due to strong winds (Thamburaj and Natarajan, 2010) (see propagation by limb cuttings).

A mass production is also the need of the hour for the fodder purpose which increases the milk production of cattle for making the milk available to the people of underdeveloped nations to meet the nutritional requirement and combat child malnutrition (Sanchez et al., 2006).

Further, it was reported that some of the *Moringa* spp. are in the process of extinction and are endangered. The preservation of *Moringa* spp is thus of great concern from biodiversity, ethno botanical, dietary and pharmacology perception (Saini et al., 2012).

The answer to all of the above appears to be in micro propagation i.e. tissue culture of *Moringa* spp. Hence, tissue culture may be an available alternative for rapid propagation of moringa (Stephenson and Fahey, 2004). Thus developing tissue culture methods for *Moringa* spp. is urgently needed. As an alternate to traditional methods of propagation, micro-propagation offers an efficient method for propagation, as it is quick, required only small amount of propagating materials (explants) and the potential to provide very large number of cloned plants. The production of micro propagated plants per unit area and per unit time is substantially higher than the traditional methods of multiplication, besides providing disease free plants.

Tissue culture techniques are based on totipotency concept of cells and play an important role in plant micro-propagation and conservation of plant germplasm (Jaskani et al., 2008).

Till to date a few attempts have been made in India and elsewhere to standardize and develop protocols for micro propagation of moringa, using different types of explants like nodal segments, stem pieces, seeds etc. (Kantharajah and Dodd, 1991; Mohan et al., 1995; Madhumathi, et al, 1995; Mughal et al., 1999; Stephenson and Fahey, 2004; Jahan and Khatun 2005; Xiang et al., 2007; Wang and Qiang, 2008; Saini et al., 2012; Shookhmand and Drew, 2013; Shahzad et al., 2014).

However, tissue culture plants of moringa are not available commercially at present for field planting (like banana tissue cultured plants) although the developed micro-propagation plants were acclimatized successfully on transfer to field condition.

CHAPTER 9

Planting

A. Field Preparation

Preparation of the field is the first step in the cultivation of crops. It depends on the type of soil, its previous history (Virgin or already under cultivation), water source, whether the intended crop is raised as rain fed or irrigated crop, farmer's plans, purpose for which the crop is cultivated (fodder, fence, leaf, pods, seeds etc.), whether intended to take up intercrops, included in agro-forestry system etc. In case of virgin land, first all the existing vegetation on the land should be removed by uprooting along with the roots. Boulders, stones etc. and particularly termite hills should be completely removed from the site. In case of field already under cultivation, nothing much may be needed in field preparation, except removal of the previous crop, its stubbles, roots etc.

In case of moringa (*Moringa oleifera* Lam) the intended field at first should be ploughed deeply and thoroughly using mould board plough in case of virgin land followed by disc ploughing to crush the soil and roots of existing vegetation. Next the field may be harrowed 3-5 times thoroughly till fine tilth is obtained. In case of field under cultivation mere ploughing 2-3 times with harrows may be enough to get fine tilth. Organic manures like FYM, compost or well decomposed poultry (chicken) manure may be applied and incorporated into the soil during the final ploughing. If necessary tank silt may be added to enhance the fertility of the soil. It is advisable to raise green manure crops; sunnhemp (*Crotalaria juncea*), daincha (*Sesbania acculeata*) etc., *in-situ* and incorporate into the soil before taking up of planting of drumstick seedlings or limb cuttings or direct sowing of seeds. *In-situ* raising and incorporation of green manures crops before flowering is necessary to enrich the soil fertility status, though moringa is not a fertility demanding crop. After incorporation of organic manures / green manure crops, field may be ploughed once and finally leveled prior to planting of moringa. This preparation of field may be started during May (summer) and continued upto June-July till the planting of moringa. In case of irrigated crop irrigation channels may be laid down before or after planting of moringa depending upon the time available. In case of rainfed crop basins may be made after planting moringa.

B. Digging and filling of the pits (holes)

On the leveled land pits are dug up in May-June, 3-4 weeks before planting of moringa. Pits (holes) will serve to loosen the soil and help to retain moisture in the root zone enabling the roots of seedlings / limb cuttings to develop rapidly and penetrate deeply. The dimensions of pits depend on the type of soil, type of moringa (annual or perennial), type of planting material (seed, seedlings or cuttings). The recommendations on pit size range anywhere from 30 cm^3 to 90 cm^3 depending on the hardiness of the soil. In loose sandy loams, small holes (45 cm^3) and in heavy soils bigger sized pits (90 cm^3) may be dug at pre-determined spacing. Further, irrespective of type of soil for planting cuttings, 90 cm^3 pits would be enough. In case of drumstick based agri-horticulture system Pande et al., (2013) advised a pit size of 75 cm^3 at 4 × 4 m spacing.

Dug out pits may be left open for weathering and drying for 7-10 days and then refilled. Pits may be filled up with a mixture of top soil plus 10 kg of FYM / compost plus 250 g of deoiled neem cake plus 250 g of single super phosphate or 100 g of diammonium phosphate (DAP). A mixture of 5 kg FYM + 100 g of ammonium sulphate + 200 g single super phosphate + 100 g potassium sulphate mixed with dugout top soil may also be used in filling the pits (Pande et al., 2013). Aldrin or BHC dust should also be applied and mixed in the pit soil against termites and root grubs. Avoid using soil from the bottom of the pits as it is not completely weathered. On the other hand, top soil of the pits is comparatively weathered and contains beneficial microorganisms that can cause more effective root growth of seedlings / cuttings. Bottom soil of pits may be utilized to form a bund around the plant basins. Again in areas of heavy rainfall, a mound may be built with dug up soil around the base of the planting material for drainage and also to avoid water coming in contact with the base of the plant. After filling, the pits may be watered for settlement of the filled soil a couple of days before planting seedlings / cuttings or wait until a good rain is received before out planting seedlings / cuttings.

C. Spacing

Generally moringa is planted in square system of planting, but sometimes rectangular system for intercropping as well as for leaves and pods is adopted as in Gujarat (Jadav et al., 2009). For live fence moringa is planted in single rows at close spacing.

Spacing between plants and rows plays an important role in maintaining adequate population, sufficient room for the growth of the plants. Determination of optimum population and space between plants and rows is the most important pre-requisite to achieve maximum yield of any crop (Jadav et al., 2009). The spacing of crop plants, including drumstick depends how they will be grown, besides type of soil, variety, type of moringa, whether rainfed or irrigated etc. Recommended spacing is anywhere from 1.2 to 3.5 m for intensive production of green matter (Doerr and Williams, 2007-09).

Drumstick is a strong demander of sunlight, hence the spacing should be decided in such a way that it gets adequate sunlight and air flow in the plantation. In view of this moringa is commonly planted at a spacing of 3 × 3 m to 5 × 5 m in block plantation for pods or at 5 m spacing in single line for live fence or hedges (Ramachandran et al., 1980; Nautiyal et al., 1987). For getting good quality leaves and pods, PKM-2 moringa should be planted at 2.5 × 2.5 m spacing (Devi et al., 2007).

If the crop is raised for production of leaves only, plants may be spaced at 50 cm × 100 cm. If using rised beds, beds may be formed with 60 cm wide tops and plants are spaced at 100 cm apart in a single row. For intensive production of leaves for fodder, plants are spaced at 10-20 cm with in rows and 30-50 cm in between the rows. Closer spacing allows harvest of young edible shoots every 2-3 weeks compared to wider spacing.

If the trees are grown to their full mature height they should be spaced at 3 × 3 m to ensure sun light and air flow. This spacing would apply to a few trees planted in home steads (Anon, 2012b). Trees are often spaced 100 cm or less apart in a line to create a live fencing post (Anon, 2012b). Live fences or hedges of drumstick are typically established at a spacing of 1 m or less and managed for foliage through frequent cuttings (Morton, 1991). In alley cropping there should be 10 m distance between rows (Beaulah et al., 2010). In agro-forestry, a spacing of 2 and 4 m is adopted in moringa elsewhere (Saint Saveur and Broin,

2010). For raising intercrops sufficient row spacing should be provided to ensure sufficient spacing and good production of both main crop and intercrop. In intercrop system often moringa is spaced at 3 × 2 m, but wider spacing is desirable, which benefits pod production at the cost of leaf production per unit area. Wider spacing promotes pod production through greater branching and flowering. Hence spacing need to be adjusted according to the intercrop to be raised (Beaulah et al., 2010). Wider row spacing provides space for easy inter cultural operations like weeding, manuring etc. (Jadav et al., 2009). Appropriate row spacing also renders scope for better growth and development of crops, which reflects in higher crop production (Jadav et al., 2009). Hence, rectangular system of planting is better than square system for intercropping.

Limb cuttings (of perennial types) are spaced at 5 × 5 m for pod production (Kumar and Pugalendhi, 2010). Singh etal., (2014) reported that one-year-old stem cuttings are planted at 3-5 m apart whereas for annuals (like PKM-1) less spacing is adopted. For pods and leaves (for human consumption) moringa should be planted at 2-5 m spacing depending on the type of moringa and type of soil (Kashyap et al., 2009). Spacing of atleast 2×3 m is recommended for perennial types depending on pruning frequency of the plants, their shade tolerance and other requirements of the companion trees as well as space required for access of equipment (Beaulah et al., 2010).

Horticulture Research Station, GKVK, University of Agricultural Sciences, Bangalore recommended spacing of 5 × 5 m for tall types and 3.2 × 3.2 m for dwarf (small) types under rainfed conditions of Karnataka (Kumar and Pugalendhi, 2010).

In Andhra Pradesh, a spacing of 2.5 × 2.5 m in square system is followed for PKM-1 variety providing 1600 plants / ha (Reddy et al., 2010). Peter (1978) and Verma and Chaarasia (1993) recommended 5 × 5 m spacing in square system for moringa in Kerala. The farmers of middle Gujarat agro-climatic zone III are advised to plant PKM-1 drumstick at a spacing of 2 × 2 m for higher yield and higher net returns (Jadav et al., 2009). In Maharashtra a spacing of 3 × 3 m is followed providing 445 plants / acre for pod production (Anon, 2012b). Janakiram et al., (2017) suggested that the spacing for perennial moringa should be 5 × 5 m, for annuals 2 to 2.5 m on either side and for high density planting (HDP) of PKM-1, 1.5 × 1 m spacing. In Philippines, moringa trees are usually spaced at 3-5 m apart, unless they are being used as a live fence, in which case they are grown much closer (Anon, 2012a).

However, a spacing of 5 × 5 m may be considered generally appropriate for moringa for most situations in respect of pod production (Anon, 2015b).

D. Planting

Planting of moringa plantation is done by three methods of propagation: (i) direct seeding in the main field, (ii) transplanting seedlings from nurseries, (iii) limb cuttings.   Under favourable conditions direct seeding is preferred. This is mostly adopted in case of annual moringa. In case of delay in land preparation, seedlings are raised in nurseries and the seedlings of 45-60 days old are transplanted in the main field at appropriate time. In case of perennial types of moringa traditionally limb cuttings are planted generally in the main field (Pande et al, 2013).

Time of planting is more important as it decides the survival and further performance of transplanted seedlings / cuttings. Generally, drumstick is planted on the onset of monsoon season between June to October under South Indian conditions. The thumb rule is to plant seedlings 7-8 months before harvest of pods as the moringa comes to flowering in January – February irrespective of date of sowing seeds. So seed sowing may be done during June to October. In the regions, where irrigation facilities are available transplanting of seedlings can be done during January – February months besides planting in July – August period. Under sub-tropical conditions of North India Singh et al., (2014) found June-July and November-December well suited for sowing or planting moringa, provided there is no freezing temperature in the latter months at Panthnagar, UttaraKhand, India.

The limb cuttings should be planted during the monsoon months of June to August to avail the benefit of monsoon rains for easy rooting of cuttings and further growth (Kumar and Pugalendhi, 2010).

It is advised to plant the seedlings late in the afternoon (after 3-4 pm) to avoid the hot sun on the first day of planting. Make a hole in the pits (the size of root ball of earth) to accommodate all the soil in the container (polyethylene bag). The bag may be cut open at the sides / bottom carefully and place the seedlings in the scooped out hole without disturbing / damaging the root system. The soil around the base of the seedlings may be pressed. The transplanted seedlings are watered immediately and thereafter lightly for ten days.   Under moderate clay conditions watering should be done just to optimum levels to avoid root rot.   However, there should not be any water stress till the establishment of seedlings / cuttings (Kumar and Pugalendhi, 2010).   Support may be provided to the newly planted seedlings to avoid falling during watering / by winds.   The seedlings may be protected from termites etc and diseases, cattle etc.

Under water logged conditions and poorly drained soils, the seedlings may be planted on mounds so that excess water is drained and the collar region of the plant does not come in contact with water.

In case of cuttings, they are planted by burrying bottom one third length of the cuttings in the pits and the soil is pressed firmly and watered. The same precautions of seedling planting may be followed in the case of cuttings.

To ensure adequate sunlight and air flow, the rows of moringa may be oriented in East-West directions.

CHAPTER 10

After (Care) Cultivation

It is often stated that the cultivation of drumstick is easy and crop does not require much care, once it has established well in the field. However, in fact moringa requires a lot of care and maintenance after planting in the field to produce the expected yield for longer time. It needs care with respect to plant architecture, nutrition, water management and health management (Singh, 2011).'

Moringa is very fast growing tree, but at the same time it is very weak and breaks easily due to winds and weight of the fruits. If left unattended and untrained it grows straight to a height of 10-12 metres and tends to produce long branches near the base of the tree and long fragile branches that grow vertically and produce leaves and fruits at their extremity, resulting in less production of leaves and fruits (pods) with a tendency for flower drop. Such tall trees also pose harvesting problem, since the leaves and fruits are out of reach of hand harvesting. On the other hand, branches that arise near the base of the tree, extend laterally and interfere in carrying intercultural operations. It is, therefore, essential to give the trees a good shape, when they are young by encouraging lateral branches and reducing height, thus creating a bush shape, more productive and facilitate easy harvesting. Further, to make good and strong framework the trees should be trained by means of pinching, training and pruning of excessive growth, so that it may be able to tolerate the weight of fruits and winds (Verma and Chaarasia, 1993; Pugalendhi et al., 2010).

Pinching (Tipping, Nipping)

Training of moringa trees should start right from the seedling's stage by pinching (nipping) of growing tips of seedlings. Pinching is a mandatory operation in PKM-1. If not done, the tree will grow excessively tall without branches and produces only few pods. This is the first operation of after cultivation practices. The process of pinching is as follows: when the seedlings establish well and reach a height of 50-100 cm in the main field (Kathiresan et al., 1999; Beaulah et al., 2010; Pugalendhi et al., 2010; Madalageri et al., 2010) the process of pinching should be initiated. Opinions differ regarding the age of the seedlings at which the pinching process should be started. Vijaya Kumar et al., (1999), found that early pinching of growing tips carried out 60 days of after seed sowing was better than pinching 90 days after seed sowing for obtaining better yields. On the other hand, Devi (2003) suggested that in case of moringa plant set at a spacing of 2.5 × 2.5 metres pinching 80 days after seed sowing was best from the stand point of growth, yield and quality of annual moringa cv. PKM-1. Since, PKM-2 produces longer pods, often touching the ground and getting bruised, a pinching height of 120 cm from the ground level was recommended in Tamil Nadu (Devi, 2003). Pinching 80 days after sowing, according to Devi et al., (2007) provides good quality pods and leaves in plants spaced at 2.5 × 2.5 metres and gives superior performance of all the pod characters in cv. PKM-2. Physiological attributes like leaf weight and light interception were at their best levels in early pinching. Early pinching was characterized by least height (2.2m) with better branching (31.3) and stem girth (2.38 cm). Similarly canopy spread, total dry matter production and root volume were distinctly superior in early pinching (Pugalendhi et al., 2010).

The process of pinching includes the nipping of the terminal growing tips of the seedlings 10 cm from the top with fingers, since the terminal growth is soft, tender, devoid of bark fiber and brittle and therefore easily broken. Alternately, sharp tools like a knife or a shear can be used. Tipping of growing tips releases the apical dominance of terminal bud and triggers the lateral branches (secondary branches) below the cut on the pinched stems in about 7-10 days of operation. When these secondary branches grow and reach a length of 20-25 cm (after 45 days) they are cut back in turn to 10 cm length, which results in the production of tertiary branches (after another 45 days) which are pinched in the same manner as in the case of secondary branches. Thus, pinching is done four times before the commencement of flowering (when the tree is about 3-4 months age) (Palada and Chang, 2003; Beaulah et al., 2010; Madalageri et al., 2010; Pugalendhi et al., 2010).

Pinching has several advantages:

- Pinching encourages the tree to become, bushy (globose) with maximum bearing canopy,

- Produces many pods within easy reach from the ground.
- Helps the tree to develop a strong framework and reduces the damage due to heavy winds and fruit load.
- Promotes the growth of many lateral branches, increases yields and reduces tree height
- Allows higher interception of sunlight,
- Allows higher flowering and higher yields
- Makes it easy to adopt intercultural operations and plant protection measures
- Helps in early initiation of flowering
- Increases yield by 16-18% than the unpinched plants (Madalageri et al., 2010).

In case, the trees are older and pinching was not carried early enough the main stem can be cut with a sharp tool, just above a node. Cuttings in the internodes will cause the rotting of the stems all the way down to the node below the cut and provides access to pests, diseases and parasites, which should be avoided (Pugalendhi et al., 2010).

The plants of cv. PKM-1 are usually trained to central leader system by pinching when the plants attain a height of 75 cm, 2 months after sowing seeds. Palada and Chang (2003) advised to trim the apical growing shoot when the tree is 1 to 2 metres height in Taiwan. For cv. Mali, it was recommended (to encourage the production of many branches and pods within reach of the ground) to cut off the central growing shoot tip when the plant is 1.5 – 2.0 metres in height and regularly cut off the tips of branches to encourage bushy shape of the tree (Doerr and Williams, 2007-09).

Pinching not only influences the growth and development of plants but also plays an important role in production and quality of pods and seeds in drumstick.

Training and Pruning

After the initial trimming (pinching) and shaping of the trees, maintenance pruning is required for moringa. Moringa is an extremely fast growing tree and becomes tall and lanky particularly in the absence of training and pruning and renders harvesting difficult, hence it is necessary to prune frequently beginning when they are still young. Young plants are to be trained to develop a strong framework of scaffold limbs to bear large crops of fruits without breaking. Pruning is an important operation and essential practice to increase yields. Pruning in drumstick is very essential because it bears pods on current season's growth and requires regular annual pruning to replace the old and unproductive wood by the new growth. The total vegetative canopy continuously enlarges year after year with the reduction in bearing area, pod size and ultimately the tree becomes weak and less productive or unproductive. Hence, to maintain tree vigour, fruit size, quality and yield regular judicious pruning is necessary (Kalalbandi et al., 2014). Pruning is essential to facilitate profuse flowering, fruiting and to keep plants in control for easy harvesting (Nigam and Mishra, 2003). Pruning is important to induce higher yields of both leaves and pods. Trees may be maintained at 2-3 metres height to give an access to leaves and pods during harvesting. By training and pruning, a clean bole (trunk) of about 1.5 to 2.0 metres length is maintained. Pruning provides better conditions for light, nutrition and moisture for plant growth, which results in timely commencement of reproductive phase, resulting in more number of fruits. For large number of flowers and sturdy fruits pruning is necessary and also in case of tall trees, where harvesting is a problem. Pruning is an important operation and essential practice to improve the morphological characters of fruits.

The method of pruning differs from end product to end product. In case of plants grown for leaves which are harvested by cutting, all the stems above certain height are cut off to encourage branching and leaf production. If the leaves are harvested by plucking or if the trees are left unharvested during the dry season, the bushy shape of the trees is lost and a good pruning must be done at the earliest of rainy season (Pugalendhi et al., 2010). In Niger, the moringa trees are cut down to 20 cm above the ground once or twice a year for increased leaf production. Trees grown to harvest leaves are to be regularly pruned to maintain steady production of leaves and to maintain the tree height of 1.0 to 1.5 metres for easy harvesting. In areas where there is commercial production of leaf powder, plants that reach a height of 1.2 meters are cut back to a height of 20 cm. This minimizes the growth of woody stem material and optimizes the leaf production. In Senegal where this is being done, there are 6-9 harvests of leaves of moringa each year (Anon, 2012c).

In seed producing farms pruning helps to induce more fruits as well as larger fruits. Ten percent of fruiting occurs on straight branches, while 90% fruiting occurs on lateral branches. Therefore, to achieve higher yield, trees are to be pruned so that more laterals are produced.

Since pruning of trees strongly influences the plant growth, flowering, fruiting, yield, the number of pods, edible portion of pods and quality of pods it is essential to define the extent of pruning (Farse et al., 2004, 2006). A pruning trial conducted in Maharashtra to study the effect of pruning indicated that under severe pruning i.e. 50-70% removal of branch growth, fruiting was delayed, but under medium pruning i.e. 25% pruning, trees flowered and fruited early with maximum yield (Farse et al., 2006). Further, under severe pruning pod number, edible portion and fruit quality were reduced, but under medium pruning, pod number, edible portion and fruit quality in terms of pulp per pod, number of seeds per pod, weight of seeds per pod, distance between two seeds in a pod and weight of single seed increased (Farse et al., 2004). Medium pruning i.e. 70 cm from the tip resulted in increased yield of pods in perennial types of moringa (Aundipatti type) (Anon, 2005).

Pruning is done after harvesting of the pods. Once the trees have stopped bearing, branches have to be cut off for new branches and fresh bearing wood. However, pruning studies in Karnataka indicated that heading back of the plants a month before monsoon (May-June) was the most suitable one (Madalageri et al., 2010). Singh et al., (2014) advised pruning after harvesting the crop at 75 cm height for achieving excessive branching and bushiness. Farmer should do atleast two prunings per year and harvesting of pods can be the occasion to start pruning.

Generally, perennial types are allowed to fruit without any pruning. It is felt, that in perennials, once they were pruned for shape, there is no need to prune for another 10-12 years, but pruning should be continuous process every year or as and when necessary or as and when required in which case, dead, damaged and unproductive twigs are pruned for sanitation of the tree. Pruning is necessary atleast once in a year in case of perennials (Anon, 2013). In case of perennial live fence, moringa trees are often cut back year after year to maintain the height of the fence. Older trees that are unproductive or too high for harvest can be pruned at ground level for rejuvenation. If the main stem is too thick, terminal and lateral branches may be cut off for rejuvenation.

To maintain a bushy appearance, the tree can be cut back annually to one metre height or less from the ground. Frequent pruning of growth tips will maintain and increase leaf growth, and there by plant height can be controlled for easy harvest. Otherwise, the tree trunk becomes lanky and less productive, difficult to harvest. Involuntary removal of tender tips for culinary purpose also results in bushiness of plants. Older flowering branches may be repeatedly pruned to stimulate production of new branches / shoots on which bearing occurs. In any case, it is important to cut just above a node to reduce rotting of terminals. Unwanted branches and twigs may be removed.

After pruning, the tree basins have to be prepared and manured preferably with organic manures. The inter spaces need to cultivated for intercropping during monsoon period. On receipt of rains, recommended levels of fertilizers (NPK) are applied (Madalageri et al., 2010).

In view of the above, it was advised that as moringa seedlings grow,a programme of consistent pruning is necessary for encouraging bushy structure and maximum production. To maintain a bushy appearance, the tree can be cut back annually to one metre height or less from ground level.

Pollarding

As a result of its straggly habit of growth old moringa trees of perennial type will often become ill shaped, and ugly with high branches at the end of which the fruits will be produced with attendant difficulties in harvest. Such trees also remain less productive or unproductive totally. In such trees, pollarding may be necessary and useful. Pollarding consists of removing such old, high, ill shaped and unproductive branches by pruning. This will not only result in a good shaped tree, but also promotes new growth on which fruiting occurs. This practice of pollarding may be done repeatedly (Muthuswamy, 1954). It is a common practice in perennial drumsticks to keep the tree in good shape and in production. The tree is cut back to about 1.2 to 1.5 metres in height to promote branching, fruiting and make it easier to harvest the leaves and pods. At some places, moringa is pollarded on a 90 days cycle to provide greens for marketing (Malik, 1992).

Perennial types are pollarded to 0.3 to 0.45 metres in height from the ground during October-November followed by the application of organic manures (25 kg/tree) plus the recommended levels of NPK fertilizers. New shoots arise from the stump. The trunks are earthed up. Several shoots sprout from the stump out of which 5-6 shoots on each plant are retained, while the rest are removed. When these shoots grow to a height of 60 cm they are pinched. Similarly the shoots that arise on the above retained shoots are pinched when they grow to 60 cm length. When this is done, trees grow dense with many productive branches. Pollarded trees flower in March and pods will be available upto September.

Ratooning

Pod yields in drumstick are not uniform in all years of their existence in the field. There will be fluctuations in the yield from one year to another year. After some years, depending upon the type / cultivation and management trees produce either uneconomical or no yields. So to obtain uniform yields in the following years a treatment called "ratooning" is adopted since long in South India. Further, due to precocity in flowering and bearing heavily and persistent flowering and fruit production without any definite peaks, trees of annual types (cv. PKM-1) become exhausted and appear senescent at the end of the year, due to loss of vigour. However, such trees do not die on their own accord, but linger weakly without production. Yields will be high during the first year of bearing. It is conceived that yields would get reduced during the following years and that the vigour and yields can be restored by the treatment "ratooning".

The treatment of ratooning consists of cutting of the main stem to about 90-100 cm from the ground level followed by manuring and other management practices. The cutting is done immediately after the harvest of pods (Beaulah et al., 2010; Pugalendhi et al., 2010).

About two weeks later, the stump produces several number of new shoots (15-20), below the cut densely (Plate 10.1). Out of these many sprouts, 4-5 robust and healthy shoots are allowed to grow further by removing others when young itself before they grow longer and harden and deprive of trees reserve food. These retained shoots when they grow to a length of 60 cm or more, bearing starts on these retained shoots by 4th and 5th month after cutting the main trunk. This is recognized as the second crop or first ratoon crop. After the harvest of this crop, the ratooning process is repeated to get a third crop i.e. second ratoon crop. After the third crop (second ratoon crop) the trees of annual types are entirely removed by uprooting and fresh moringa plantation is raised by planting seed directly or planting new seedlings from nurseries. Generally 2-3 ratoon crops are advisable and allowed.

Fig. 10.1 Regrowth on a stump of moringa for rationing.

Annual types are raised both under adequate and inadequate irrigation systems. So the process of ratooning differs slightly between them. In case of trees under adequate irrigation situation, the tree trunk is cut back (foundation pruning) to 100 cm from ground level and new shoots are allowed to grow to produce pods. Such new growth is copiously irrigated to revive the vigour and growth. Too many (often more than 15) new shoots appear densely on the stump below the cut, of which 3-4 shoots are retained to grow as primary branches by removing the rest of the shoots at tender stage itself (Palada and Chang, 2003).

The south-west monsoon is not assured every year fully and during the month of June to August often dry-spell prevails which does not support apical re-growth. Under these conditions moringa is grown under inadequate water conditions. It results in wilting of whole crop without foundation pruning and fresh shoots emerge all over the canopy. After the receipt of rains the whole canopy becomes hardier year after year. The pod size and length of pod become smaller than the size of the pods from regularly pruned and irrigated moringa crop. At this stage to restore yields and quality of pods the trees are subjected to the 'ratooning' treatment. In this system four ratoons are possible after which the crop is removed from the field and new crop is raised freshly (Beaulah et al., 2010).

In case of perennial types of moringa, a more or less similar type of ratooning is followed once in 4 or 5 years (not continuously year after year as in case of annual types) in which trees are cut back to one metre height from the ground level and regrowth is allowed on which flowering and fruiting occur (Beaulah et al., 2010). However, under normal conditions perennial moringa trees will live for 15-20 year after which it may not be economical to retain the crop in the field. Therefore, for continuous production, it is preferable to replant the field by fresh limb cuttings, which are planted in the interspaces of existing trees. However, this should be done before they become unthrifty and uneconomical (Beaulah et al., 2010).

After every cutting of the trunk to a stump the plants are manured and fertilized with recommended doses of manures and NPK fertilizers followed by irrigation. The cut ends may be pasted with copper oxy chloride (3 g/l) or sprayed to protect the cut ends from pests and diseases.

Thus, many types of drumstick are grown for 3-4 years and then replaced with new stock on a regular rotation, almost as if they are biennials.

De Blossoming

In case of moringa trees grown for pod and seed production it has been recommended to remove the flowers of the first year, as this blossoming increases the pod yield in the following year. Removal of flowers will channel all of the young plant's energy into vegetative and root development (rather than energy-draining pods) leading to more vigorous vegetative growth for the survival and establishment of young plants and pod yields in the future. (Palada and Chang, 2003).

CHAPTER 11

Manuring and Fertilization

Moringa (*Moringa oleifera* Lam) trees generally grow well in most soils without manuring and fertilization (Palada and Chang, 2003). Muthuswamy (1954) stated that manuring of moringa (perennial types) was seldom practiced in India as the tree was hardy and capable of growing in poor soils once established, because of its extensive and deep root system, which is efficient in foraging the nutrients from the deeper layers of soil (Palada and Chang, 2003). However, for optimum growth and yields, manuring and fertilization of moringa plants are necessary and important. According to Seemanthini (1964) application of manures promotes larger yields. Martin and Rubertie, (1975) stated that usually moringa trees do not require any horticultural attention and opined that mulching and fertilization would provide better growth and results in better quality of its products. According to several moringa research workers and farmers, fertilization and manuring are important factors in augmenting the performance in terms of yield (Hanchinamani and Madalageri, 1994). As drumstick is a perennial tree, for maintaining production in each year, optimum fertilizer management is essential to sustain as well as enhance growth and production of plants. The use of optimum levels of manures and fertilizers is an important factor in improving the crop yields.

Though moringa can be grown in backyards, on field bunds etc., as sole or group of trees without manuring, but in commercial cultivation manuring and fertilization are required, Moringa crop ordinarily does not require more manuring and fertilization if only pods are desired, but for commercial cultivation of leaves, fertilization is necessary, because leaves are important for making powder and to use as animal feed, hence it is a must and necessary input. Despite this, *Moringa oleifera*, is not very much demanding in terms of fertilization, but a minimal application of manures improves its growth and pod yield.

Generally, the manures and fertilizers, mixed with top soil of pits applied while filling the pits (holes) would be enough / sufficient for one year. Thereafter manuring is necessary, so 10 kg of FYM per tree per year is applied in a ring dug around the tree in the basin. Afterwards it is important to apply manures and / or fertilizers at least once in a year, for instance before the rainy season, when the trees are about to start an intense growth period (Saint Sauveur and Broin, 2010).

Moringa is grown in several states of India and other countries as well with much differences in agro-climatic conditions and genotypes (varieties) wherein the cultivation practices and the purposes for which it is cultivated differ. In African countries, Philippines, Pacific region etc., moringa is largely cultivated for its foliage to feed malnutrition affected children and lactic mothers and cattle. But in India, moringa is mainly cultivated for its fruits (pods), but often, for its seeds both for propagation, oil extraction and biofuel and export. In the following pages the details of manuring and fertilization are provided for moringa grown mainly for pods.

Moringa requires manures and fertilizers at different stages of its growth, and development viz., at sowing and growing in nurseries, at planting time, during initial establishment and growth, pinching and pruning, flower initiation, first flowering, fruitset and fruit growth and development. Accordingly the kind of manures and fertilizers differs. Further, the doses of manures and their requirement differ from growing region to region depending on the fertility status of the soil; similarly the sources and nutrients differ. Further, the suggestions / recommendations seem to differ, which are mostly arbitrary and not based on experimental basis except in few cases.

During nursery period, the nutrients in soil should be high enough to sustain the plant life cycle (Wilson et al., 2001) and can be periodically increased with doses of organic manures and chemical fertilizers like NPK (Beaulah et al., 2004). To improve the nursery production of *Moringa oleifera* seedlings the use of small amounts of compost can be recommended, because compost positively impacts plantlets providing 100% survival rate. The efficiency of compost is attributed to its average mineral composition, because high amount of nutrients has negative influence on the survival rate of plantlets (Haou Vang et al., 2017). Cow dung (100g) and poultry manure (100g) showed a positive impact on plantlets survival rate (100%) (Haou Vang et al., 2017).

Research work done in Tamil Nadu (India) indicated that application of 75 kg FYM + 0.37 kg of ammonium sulphate per tree in December gave threefold increase in yield over un-manured trees (Sundararaj et al., 1970). Kader and Shanmugavelu (1982) reported that application of 20 kg FYM per pit, top dressed with 100 g each of urea, super phosphate and muriate of potash per tree gave higher yield. According to Sunthanapandian et al., (1989) moringa requires NPK at 44: 16: 30 g respectively per tree at the time of first pinching (75 days after sowing) 44 g nitrogen per tree at first flowering (150 days after sowing) as top dressing in PKM-1 cultivar to get higher yield. The trees should be fertilized with 100 g urea, 100 g super phosphate and 50 g muriate of potash at third month of planting and again at sixth month of planting 100 g urea per plant should be top dressed (Ramachandran et al., 1980), Kathiresan et al., (1999) suggested to apply urea (100 g), single super phosphate (100 g) and muriate of potash (50 g) and gypsum (50 g) followed by irrigation three months after planting in Tamil Nadu (India).

Moringa oleifera is generally grown successfully without fertilizer application in Kerala. However, green leaves, FYM and ash are put in the trenches dug around the tree about 60-90 cm from the tree base during rains and covered with dug up soil (Rai and Yadav, 2005). To get higher yields from moringa, trees should be fertilized with urea (400 g) + single superphosphate (100 g) + muriate of potash (50 g) per tree at third month and at sixth month of planting urea (100 g) should be top dressed (Subramanian et al., 1998). In Andhra Pradesh, after 45 days of planting, 40 g of nitrogen (N), 15 g phosphorus (P) and 30 g of potassium (K) are applied per tree followed by 45 g N, 75 days after planting with 500 g FYM and 250 g neem cake (Reddy et al., 2010). To get higher yields in annual moringa pits may be filled with 6 kg FYM plus dug up top soil + 200 g neem cake, 100 g urea, 100 g super phosphate and 75 g muriate of potash per plant after third month of planting; after the start of cropping, the same quantities of manures and fertilizers as top dressing in the second year after pruning after every third month and after the start of cropping again, the same above quantitates of manures and fertilizers are applied (Prasant and Sreenivasa Rao, 2004). Application of 10 kg FYM plus 250 g neem cake plus 250 g single super phosphate per pit may be applied. Next after third, sixth and ninth month of planting 100 g urea plus 75 g muriate of potash per plant are applied as top dressing followed by copious irrigation (Madhusudan Reddy, 1999). Save Indian Farmers, a US based organization, supporting moringa cultivation in Maharashtra and Andhra Pradesh recommended 20 tonnes FYM per ha and each pit with 8-10 kg FYM.

According to Valia et al., (1993 a, b) moringa trees can be fertilized with 10-12 kg FYM plus 60 g N, 80 g P and 40 g K per pit as basal application in Maharashtra. Again in Maharashtra (India), in rainfed areas 200 g fertilizer mixture (components of fertilizers mixture not indicated) given in September and December, while in irrigated areas 20 kg FYM and 250 g of fertilizer mixture are given in April. Fertilizer dose is increased at 500 g per year as per the need of the crop (Anon, 2012b).

Nutrition trials at Bagalkot (Karnataka) revealed that 200:100:100 g NPK per plant per year was most optimum for moringa crop applied as basal dress before commencement of monsoon (Madalageri et al., 2010).

An investigation on the nutrient requirement of one-year-old moringa cultivars revealed that highest number (117) and weight of (5.24 kg) of pods per plant were obtained in selection 6/4 with the fertilizer combination of 250: 125:125 g NPK per plant at Dharwad, Karnataka state (Hanchinamani and Madalageri, 1994).

Based on three year trials, the farmers of Gujarat Agro-climatic zone III were advised to apply 100 g N, 25 g K and 10 kg FYM per plant in two split doses. First half dose of N, full dose of K and FYM should be given in the first week of May after planting and the remaining half dose of N in October for PKM-1 moringa for obtaining higher yields (Jadav et al., 2010).

Pande et al., (2013) reported that in Madhya Pradesh (India) farmers of moringa apply 100 g ammonium sulphate, 300 g super phosphate and 100 g potassium sulphate per tree during July every year after first year of planting.

For commercial cultivation of drumstick it was suggested to apply well rotten FYM @ 15-20 kg plus 45 g N, 10 g P_2O_5 plus 30 g K_2O per plant, three months after sowing at Pantnagar, Uttarakhand (India). After six months of planting a dose of 45 g N along with 75 kg well rotten FYM every year in June (onset of monsoon) at one metre distance from the stem in a ring should be applied (Nigam and Mishra, 2003).

Application of 10-12 kg well rotten FYM per tree per year followed by irrigation after pollarding the trees was found to be beneficial to encourage vegetative growth of moringa (Saroja and Dwivedi, 1993).

Palada and Chang (2003) suggested 500 g of commercial nitrogenous fertilizer per tree, and in the absence of nitrogen fertilizer to use compost or well rotten FYM @ 1-2 kg per tree in a trench in Taiwan. In Mali, compost is applied and found to increase yields, help trees to grow faster and develop healtheir trees that will last better the drier months (Doerr and Williams 2007-2009).

In India, the following fertilizer schedule is adopted in moringa.

1. If only organic – 10 kg FYM or compost as basal application at sowing + 20 kg per tree at pruning, biofertilizer @ 20 g / litre of *Azosirillum* and phosphate bacteria can be added.

2. If only chemical fertilizer – 44:60:30 g NPK / tree at the time of pruning (75 days after sowing) nitrogen @ 44 g per tree at first flowering.

3. In case of INM: 150 160: 100 g NPK per tree as basal application plus 500 g poultry manure, 250 g neem cake per pit.

Often deficiencies of micronutrients viz., iron and copper were observed on moringa, affecting seriously its growth and performance. Iron deficiency was observed when rains were more and drainage is poor. In affected trees leaves become pale and whitish obviously due to lack chlorophyll synthesis. The deficiency of iron may be corrected by spraying 5 g of ferrous sulphate plus one gram of lime salt twice or thrice at weekly intervals (Prasant and Sreenivasa Rao 2004). Copper deficiency occurs when moringa is grown in calcareous soils. The symptoms of copper deficiency include inward curling of leaves especially in young leaves, crinkling of leaves and in extreme cases leaves get shriveled and affected plant appears weak and sickly. Deficiency becomes acute when there is failure of monsoon rains. Copper deficiency can be corrected by spraying 400 g copper sulphate in 200 litres of water and also growing soybean as intercrop. Avoidance of incidence of copper deficiency in moringa is possible by avoiding calcareous soils for growing moringa (Prasant and Sreenivasa Rao, 2004).

Inorganic Fertilizers

Inorganic fertilizers (mineral / chemical / synthetic fertilizers) are manufactured in factories. They are generally water soluble and contain one or more nutrients. They release quickly the nutrients for plant uptake after the application. Plants respond fast by showing increased growth and production. They are easy to apply and do not decompose as the organic manure. They are relatively more expensive.

Inorganic fertilizers exhibit certain harmful effects like reduction in seed germination rate and higher seedling mortality, although mineral components uptake in young moringa plants was considerably improved. On the other hand, organic manures improve seed germination, biomass and mineral nutrition of moringa plants (Haou Vang et al., 2017).

Out of the 16 essential plant nutrients, the three major nutrients viz., nitrogen (N), phosphorus (P) and potassium (K) were experimented on moringa than the rest of the nutrients. In fact, there are almost no studies on the requirement of secondary and micro nutrients and their influence on different aspects of moringa. Hence, in view of this situation a brief account on the roles of NPK is given below:

Nitrogen

Nitrogen is an essential nutrient for crop growth, yield and quality. It plays an essential role in plant productivity, as an essential constituent of active compounds such as chlorophyll, proteins, enzymes and nucleic acid (Marshner 1995). It is also of major importance in cell division. It improves the photosynthetic process which results in higher accumulation of dry matter in plant tissues, there by enhances biomass (Zaki et al., 2012; Saleh et al., 2016).

Phosphorus

Phosphorus is the second most important nutrient after nitrogen. It is vital for photosynthesis, signal transduction, biomolecular synthesis and metabolic processes. It is required for root development. It is present in the soil as both organic and inorganic forms but is obsolete for plant uptake as it is present in immobilized, precipitated and insoluble form (Sridevi et al., 2007).

Potassium

Potassium is the third most important plant nutrient and is most abundant in higher plants. It is required in large quantities by plants. Its function is mainly osmo regulation, maintenance of electro-chemical equilibria in cells and its components and the regulation of enzyme activities (Hsiao and Lauchli, 1986). Potassium has a crucial role in the energy status of the plant, translocation and storage of assimilates and maintenance of tissue water relations (Hsiao and Lauchli, 1986). Potassium plays a key role in the crop production and its quality by improving fruit size and stimulating root growth (Saleh et al., 2016). It is necessary for the translocation of sugars and formation of carbohydrates (Zaki et al., 2015). It is essential for synthesis of protein and it mediates osmo regulation during cell division and expansion, stomata conductance and tropisms (Marschner, 1995). Furthermore, potassium is necessary for phloem solute transport and for the maintenance of certain anion balance in the cytosole as well as in the vacuole. Plants need potassium for growth and disease resistance (Chaves et al., 2005).

Organic Manures

Prior to the advent of chemicals i.e. synthetic fertilizers, farmers in many countries used organic materials in agriculture. After the advent of chemical fertilizers, particularly, after green revolution in India, there was a shift from organic manures to chemical fertilizers, because of quick results of their application and increased yields. Green revolution during 1960s and there after was dominated by the use of chemical fertilizers, high yielding varieties, irrigation and synthetic pesticides. However, soon the harmful effects of using chemical fertilizers were realized and now there is a trend to switch over back to organic manures and biofertilizers.

Several studies have demonstrated the benefits of using organic manures over chemical fertilizers in crop production of both agriculture and horticulture:

Organic manures are made up of organic wastes (plant residues or animals) (see types of organic manures). They serve as source of plant nutrients, organic matter (humus). They serve not only as source of plant nutrient but also as soil conditioners by improving soil physical properties as evidenced by increased water infiltration, water holding capacity, aeration, permeability, soil aggregation and rooting depth (Allison, 1973). Continuous and regular return of organic waste (plant remains) contributes greatly to the overall maintenance of soil fertility and productivity and reduces the need for mineral fertilizers (Parr and Colacicco, 1987). According to Stofella et al., (1997) compost and other organic manures serve as soil amendments to improve soil nutrient levels, and growth of crops, and act as a ready source of carbon and nitrogen for soil micro-organisms. Organic based fertilizers such as organo-minerals improve soil structure, reduce soil erosion, lower the soil temperature and increase water holding capacity of the soil. They improve soil water retention, stabilize soil pH, increase soil organic matter, facilitate seed germination and ultimately the growth and production of plants (Roe et al., 1997). Organic manures release many important plant nutrients into the soil and nourish soil organisms which in turn slowly and steadily make minerals available to the plants (Erin, 2007). Organic amendments are sustainable, relatively cheap materials of plant and animal origin, that are incorporated in the soil prior to seeding / planting to increase the production and the yield of crops (Ghori and Anusuya, 2017). Low cost and ready availability of materials for the preparation, gradual release of plant nutrients, ability of the treated soil to hold nutrients, being environmentally friendly and absence of harm full side effects made organic manures popular among the farmers (Adebayo et al., 2017). Continuous use of inorganic fertilizers in large doses poses environmental pollution threats and would induce imbalance of nutrients and deficiencies that culminate in poor soil quality and reduced crop yield. Besides, the rising costs of chemical fertilizers, beyond the reach of resource poor farmers also favoured the preference for organic manures (Agyenin-Boateng et al., 2006).

Organic manures are slow releasing fertilizers and usually have residual effects on the crops in subsequent cropping system.

Types of Organic Manures

Farm Yard Manure (FYM)

It refers to the refuse from animals, mainly cattle, sheep, goats and poultry. It is one of the oldest manure known and highly valued for its many beneficial effects when applied to farm soils viz., maintain soil physical and chemical properties and fertility. It consists of a mixture of four components viz., dung, litter (bedding material) waste fodder left over in the cattle shed and urine. On an average FYM contains 0.5% N, 0.25% P_2O_5 and 0.5% K_2O. A ton of FYM will supply 5 kg N, 2.5 kg P_2O_5 and 5 kg K_2O. However, the nutrient content of FYM varies with source materials. Fully decomposed FYM should be applied because it takes a year for decomposition. It is applied as basal dressing during land preparation and thereafter annually in free basins. It is applied in pits before seeding / planting seedlings / cuttings.

Compost

It refers to the material collected from the farms, house wastes from human habitats (towns, villages etc.) and allowed to decompose for a considerable period of time. It is the product of plant and animal wastes with various additions. The chemical composition of compost also varies with source material.

Green Manures / Green Leaf Manures / Green Manuring

Green manure / green leaf manure is the fresh plant material which is ploughed into the soil for the purpose of incorporating organic matter, thus supplying humus as well as the nutrients contained in the plant tissues. Generally a green manure crop is grown in the intended field before sowing /planting and incorporating into the same field and this is referred to "green manuring". Legume crops are generally grown in the field and incorporated in the same field in India. But in other countries, both legumes and non-legumes are grown and the mixture is preferred for application. The following are the commonly grown legumes in India and their manurial values (NPK composition).

Table 11.1 Commonly grown green manure crops and their nutrient contents in India (Ponnuswami, 2012)

S.No.	Name of the green manure crop	Yield (t/ha)	Nutrient content (on dry weight basis)		
			N	P	K
1.	Sunnhemp (*Crotalaria juncea*)	15-25	2.3	0.5	1.8
2.	Daincha (*Sesbania aculeate*)	10-20	3.5	0.6	1.2
3.	Sesbania (*Sesbanania speciosa*)	25-30	2.8	0.4	1.4
4.	Pilli pesara (*Phaseolus trilobus*)	8-10	2.9	0.6	1.1
5.	Cluster beans (*Cymopsis tetragonaloba*)	15	1.9	0.4	0.9
6.	Kolinchi (*Tephrosa purpurea*)	-	2.8	0.4	1.2

In some cases, green plant material including the above and others are brought from outside and applied in the field and incorporated. Such practice is called "green leaf manuring".

Oil Cakes

Oil seeds (e.g. groundnut, sun flower, sesame, cotton, neem, mahua etc.,) are rich in plant nutrients. The residue (cake) obtained after extraction of oil is rich in nitrogen and considerable amounts of N, P, K etc. It is one of the old practices to apply both edible and non-edible oil cakes as a manure. Out of the two, edible oil cakes are more profitable as cattle feeds, hence, not advisable as manures. On the other hand, non-edible oilcakes such as neem, mahua etc., are generally used as manures. The percentage of nitrogen ranges from 2.5 in mahua to 7.9 in decorticated safflower cakes. The P_2O_5 content in oil cakes varies from 0.8 to 3.0% and K_2O from 1.2 to 2.2%. Oil cakes though insoluble in water, are quick acting organic manures, their nitrogen becoming quickly available to the plants in a week or ten days after application. The solvent extracted oil cakes are somewhat more quick acting than ghani- hydraulic or expeller – pressed oil cakes. However, the quantity of organic matter added in normal application of oil cakes is too small to cause improvement in physical properties of soil. Moreover, they are somewhat costlier than other organic manure like FYM or compost. Neem cake helps in the inhibition of nitrification, losses of nitrogen in the soil, besides acting as a repellent and controlling agent of insect pests.

Poultry Manure

Poultry manure derived from the droppings, urine of birds, along with liquid excreta is a rich organic manure. It is a rich source of major nutrients required for optimum plant growth such as NPK. When applied to soil it decomposes and releases nutrients (N, P, K, Ca and Mg) significantly in the soil (Amanullah et al., 2010, Adebayao et al., 2017) and considerable soil organic matter, and exchangeable cations (Amanullah et al., 2010). It is a cheap source of nutrients and contains more nitrogen, small amounts of urea and ammonium. Poultry manure can be applied to the soil directly as soon as possible. After application it should be worked into the soil. If the droppings come from the cages or dropping pits, super phosphate may be added to these @ 1 kg per day per 100 birds to improve the fertilizer quality and to control odour and flies. It ferments very quickly. If left exposed, it may lose upto 50% of its nitrogen within a month (Ponnuswami, 2012).

Advantages of Poultry Manure (Panda et al., 2008)

1. Supplies all nutrients in right proportion to crops

2. Improves soil condition and there by retains soil fertility

3. Supplies macro-and micro-nutrients to soil as well as to crops at a cheaper cost.

4. Favours growth and multiplication of earth worms which in turn enrich the soil.

5. In general the basic nutrients in poultry manure (NPK) are the same as in commercial fertilizers and equally effective in promoting plant growth.

6. It increases moisture holding capacity of the soil and improves lateral water movement. Thus improving the efficiency of irrigation and decreasing the general droughtness of soil, improves soil retention and uptake of plant nutrients (Amanullah et al., 2010).

7. It increases the number and diversity of soil micro-organisms particularly in sandy soils. This enhances crop health by increasing

water and nutrient availability as well as suppressing natural levels of plant parasitic nematodes, fungi and bacteria (Amanullah et al., 2010).

8. It is environmental friendly and recycles waste into valuable products.

Because of above advantages it has been successfully used for crop production and was concluded that poultry manure is superior to chemical fertilizer.

Sheep and Goat Manure

The droppings of sheep and goats make very good manure. Penning of goats and sheep is, therefore, a common practice of ensuring the use of sheep and goats droppings in the fields for improving the soil fertility and thereby crop production. Sheep and goat manure contains 3% N, 1% P_2O_5 and 2% K_2O.

Besides sheep and goats, ducks and pigs are also some times penned in field to enrich the soil fertility.

Vermicompost

Vermicompost is an eco-friendly non-toxic organic manure. It is a recycled biological product from market waste (fruit, vegetables, flowers etc.) and hotels by earth worms, amended with micro-organisms like potassium mobilizer, phosphate solubilizing bacteria and free living nitrogen fixers. It requires low energy for composting waste organic materials.

It is the organic fertilizer prepared from the excrements of earth worms which is rich in organic carbon content (47%) and humus substances which help the soil structure and stimulating plant growth particularly that of roots. It can be applied to moringa along with biofertilizers and other organic manures.

Vijaya Kumar et al., (2012) stated that the sludge left over from the water after treatment can also be used as a bio-compost biofertilizer which has shown to increase yields of moringa trees.

Biofertilizers

Bio-fertilizers are carrier based preparations, containing live or latent cells of efficient strains of the microorganisms, used for seed / plant or soil application with the aim of increasing the number of such microorganisms in the soil or rhizosphere and consequently improving the extent of micro-biologically fixed nitrogen and other nutrients for the plant growth and development and finally production (Sharma and Srivastava, 2008). According to Hazarika and Ansari (2007), bio-fertilizers contain *in-vitro* cultured live or latent microorganisms capable of nitrogen fixation, zinc and phosphorus solubilization and potassium mobilization. They colonize the rhizosphere soil and help in the mobilization of nutrients for plants. They also colonize interior of the plant and promote growth by increasing the availability of primary nutrients to the host plant. Biofertilizers add nutrients through natural process of nitrogen fixing, solubilization of phosphorus and stimulating plant growth through the synthesis of growth promoting substances (Ponnuswami, 2012). Biofertilizers on application increase certain microorganisms in the soil which can change unavailable forms of nutrients into available forms that can easily be assimilated by plants (SubbaRao, 2001). For instance, most of nitrogen in the soils is tied into soil organic matter. Even after fertilization plants have to compete with soil microbes for easily available soluble nitrogen. Similar is the case with soil phosphorus. In acid soils, even when added in substantial quantities as fertilizer, phosphorus precipitates as calcium phosphate and becomes unavailable to the plants. Hence, bio-fertilizers are expected to come to the rescue and reduce the use of chemical fertilizers and restore the natural nutrient cycle of the soil. Through the use of bio-fertilizers, healthy plants can be grown, while enhancing the sustainability and health of the soil (Ponnuswami, 2012).

Importance and contribution of Bio-fertilizers in Agriculture

1. Biofertilizers play a major role in increasing the availability of nutrients, thereby improving the nutrient supplies to several crops by enhancing the humus in the soil.

2. Biofertilizers supplement fertilizer supplies for meeting the nutrient needs of crops.

3. They improve soil physical properties and soil health in general.

4. They liberate growth promoting substances and vitamins and help to maintain soil fertility and crop growth.

5. They suppress the incidence of pathogens and control the diseases.

6. They increase crop yields and result in the sustainability of the farming system.

7. They are based on renewable energy sources and are cheaper.

8. Biofertilizers are eco-friendly and pollution free, because they do not contain chemicals.

Bio-fertilizers for Moringa

There are different types of biofertilizers like nitrogen fixing bacteria, phosphate solubilizing bacteria, potassium mobilizing microorganisms, vesicular arbuscular mycorrhiza etc. However, the following types of bio-fertilizers were studied on moringa.

Nitrogen Fixing Bacteria

Atmospheric nitrogen is approximately 80% in the air. Although abundant and ubiquitous in the air, nitrogen is the most limiting nutrient to plant growth, because the atmospheric nitrogen is not available for plant uptake and plants cannot absorb or utilize atmospheric nitrogen. Some microorganisms like bacteria are capable of nitrogen fixation from the atmospheric nitrogen pool and supply to the plants in amicable form. The amount of nitrogen fixed by different systems is considerable although variation resulting from environmental conditions or different plant microbe combination is vast. The close proximity of these micro-organisms to their host plants allows efficient plant use of fixed nitrogen and minimizes volatilization, leaching and dinitrification.

The genus *Azospirillum* possesses a great potential as a general root colonizer, it contains 109 cells per gram of the inoculant. It fixes atmospheric nitrogen and makes it available to plants in associate symbiotic manner. The yield increases can be substantial, up to 30%, but generally ranges from 5 to 50%. These yield increases by *Azospirillum* are probably a result of the production of growth-promoting substances and nitrogen fixation (Ponnuswami, 2012).

Azospirillum is associative microaerophyllic nitrogen fixer, secretes mucilage and is suitable for tropical and saline, alkaline conditions. They enhance Leghemoglobin content and nitrogenous activity. They also release some plant protection chemicals (Sharma and Srivastava, 2008). It has low energy requirement but requires abundant establishment in the roots for nitrogen fixation. It can fix about 20-25 kg N/ha/season. It is used as seed treatment as well as soil application. Alternately the roots of seedlings can be dipped in the culture slurry of *Azospirillum* prior to transplanting. These are carrier based inoculants, made into slurry with sterilized FYM+soil, mixed uniformly with seeds, dried in shade and sown (Anon, 2010b). Treatment of moringa seeds with *Azospirillum* cultures @ 100 g per 625 g of seeds before sowing results in early germination and increase in seedling vigour, growth and yield.

Phosphate Solubilizing Bacteria (PSB)

Phosphorus is the most important plant nutrient after nitrogen. However soils of India are poor to medium in available phosphorus status. But most of this soil phosphorus pool is not in forms available for plant uptake. Further, the efficiency of utilization of applied phosphotic fertilizer is very low (20-25%) due to chemical fixation in soil. Besides, native soil phosphorus is mostly unavailable to crops because of its low solubility. Further, there is built up of insoluble phosphates where phosphotic fertilizers have been applied over long periods. Hence, there is need to make insoluble phosphorus into soluble and available form for plant uptake. Some hetero-trophic bacteria and fungi are known to have the ability to solubilize inorganic phosphorus from insoluble source.

The important phosphate solubilizing organisms are *Pseudomons straita, Bacillus polymxa, Aspergillus* spp., and *Pencilium degitatum* and they are referred to phosphate solubilizing bacteria (PSB). Phosphorus solubilizing fungal population is generally found more in acid to neutral soils, whereas the bacterial population is found in neutral soils. PSB are of greatest value in allowing the use of cheaper phosphorus source (e.g., rock phosphate). These bacteria grow on sulphur in soil through the production of organic acids, protons, hydroxyl ions and extra cellular enzymes. These compounds convert inorganic phosphates like, tricalcium phosphate, rock phosphate into monobasic (H_2PO_4) and dibasic (HPO_4) ions which can be absorbed by the plants (Kaur and

Kaur, 2018). These microorganisms can grow on insoluble phosphate sources such as calcium phosphate, ferric aluminum and magnesium phosphate, rock phosphate and bone meal and convert them into soluble forms. These organisms secrete various organic and inorganic acids, which act on insoluble phosphates and convert them into soluble phosphate in the rhizosphere. PSB help in the formation of chelation of hydroxyl acids with Ca, Fe and Mg and help in the production of fungi static and growth promoting substances (Sharma and Srivastava, 2008). Addition of organic manures helps in increasing the solubilizing power of the microorganisms (Anon, 2010b). Pre-treatment of moringa seeds with phospho-bacteria helps to reduce fertilizer requirement of the crops and increase the yield.

Mycorrhizae

The term 'Mycorrhiza" means fungus root, coined more than 100 years back and indicates the symbiotic association between plant roots and fungal mycelium. The hyphae of the fungus enter the cells of root (inter culture) cortex and actually provide extended absorbing surface for the invaded host plants. They are non-septate fungi and commonly known as "vesicular arbuscular mycorrhiza" (VAM), but they are now shortened as arbuscular mycorriza (AM) since vesicles are not formed in all the cases. They are characterized by the presence of special structures such as vesicles and arbuscules, hence the name "VAM". Vesicles are sac-like structures containing stored food materials chiefly carbohydrates from the host and act as temporary storage organs. Vesicles also serve as reproductive structures. Arbuscules are finely branched structures and are formed inside the cortical cells and act as extra and extended absorbing surface of the host roots. They act in reverse manner by releasing the nutrients. Mycorrhizae are ubiquitous soil borne fungi and form symbiotic association with about 90% of the terrestrial plant species. In this symbiosis, plants provide carbon source to the fungi and in turn fungi mobilize minerals (P, Zn, Cu etc.) and metabolites from the soil to the host plant to sustain biotic (root diseases and nematodes) and abiotic stresses (nutrient deficiencies, drought, chilling, salinity) conditions. AM can be inoculated through soil so as to increase the absortive surface of the roots of host plant and thereby increase the uptake of nutrients (P, N, Ca, Mg, Zn, Cu and Na). AM are obligate endo -symbionts (Anon, 2010b). AM directly absorb phosphorus through hyphae from soil and make it available to the host plant, which is otherwise in-accessible to the plants (Kaur and Kaur, 2018). The following benefits are conferred by AM fungi on the host (Kumar and Yadav, 2007).

- Enhance acquision (uptake) of mineral nutrients, especially phosphorus, which is relatively immobile in the soil

- Enhance water transport in plants, decrease transplant injury and promote establishment of plants. Mycorrtizal plants recover better following moisture stress, more efficient in water use and frequently have higher root shoot ratio than non-mycorrhizal plants.

- Reduce the vulnerability to diseases caused by soil borne pathogens.

- Increase plant growth by 20-30% in terms of biomass.

VAM differ from PSB in having two special structures viz., vesicles and arbuscules. They improve overall plant growth by improving uptake of phosphorus. Also help in stimulation of nitrogen fixation in inoculated plants and increase the uptake of nitrogen from the soil; increase disease resistance, tolerance to drought and salinity, also help in production of growth regulators and help in improvement of soil structure (Sharma and Srivastava, 2008).

Moringa plants are more productive when they are well colonized by AM fungi. The AM symbiosis increases the uptake of phosphorus and micronutrients and the growth of host moringa plant. It also improves soil quality by having a direct influence on soil aggregation and therefore, aeration and water dynamics (Ponnuswami, 2012).

Panchagavya

Panchagavya (Panchakavya) is an organic preparation made from the following ingredients:

Cow dung	-	7 kg
Cow urine	-	10 litres
Water	-	10 litres

These ingredients are mixed and kept separately in earthen pots. After 15 days, the following ingredients are added:

Cow milk	-	3 litres
Cow curd	-	2 litres
Cow ghee	-	1 litre

Tender coconut water   -   3 litres

Jaggery   -   5 kg

Banana   -   12 Nos. (Dozen)

The ingredients are mixed thoroughly both in the morning and evening and within 25 days the panchakavya is ready for use as spray. The general recommendation is, 3% solution (3 ml in 100 ml) for spraying at monthly intervals to crop plants. Panchagavya contains effective micro-organisms which in turn increase the yield and resistance to pests and diseases of crop plants.

Methods of application of Bio-fertilizers

Biofertilizers (microorganisms) are already present in the soil, but in insufficient quantities to show cognizable effects. Hence, biofertilizers have to be properly applied to the seed, seedlings or to the soil for colonization or effecting significantly. The methods of application, however, differ with the kind of biofertilizers (Sharma and Srivastava, 2008).

Azosprillum

A slurry of the *Azosprillum* is made by using sugar in water (not if adhesive (gum) is added).Then required amount of seeds is added to the inoculum slurry and dried in shade and sown immediately.

Phosphorus Solubilizing Bacteria (PSB)

For the application of PSB, the best method is seed treatment. Other methods like seedling treatment and soil treatment are also practiced.

Seed Treatment: A slurry of 200 g of the biofertilizer is made in 200-500 ml of water. Then slurry is poured slowly on 10-25 kg of seeds and seeds are dried in shade and sown immediately.

Seedling treatment: A suspension of 1-2 kg biofertilizer is made in 10-15 litres of water and the roots of the seedlings are dipped in the suspension for 20-30 minutes enabling them to absorb.

Mycorrhizal Application (Kumar and Yadav, 2007)

For successful mycorrhizal infection, the inoculum is placed in the root zone of actively growing plants by one of the following methods.

Banding or side dressing: In this method AM inoculum is placed to sides of seeds or seedlings in the zone of root penetration / proliferation.

Mixing soil with inoculum: This is the most natural method for inoculating plants. However, it requires large amount of inoculum for rapid infection. Generally this method is used for inoculating plant in nursery.

Seed pelleting: In this method, inoculum is extracted from a pot culture by wet sieving and decantation, roots retained on sieve are fragmented and added to the inoculum. Five ml of aqueous solution (1% w/v) of methyl cellulose and calcium carbonate are mixed with 35 ml of mycorrhizal sievings. This material is poured over seeds (amount depending upon the crop species), mixed thoroughly and seeds are dried and sown.

Pre-inoculation of seedlings: In this method, inoculum is placed in the seed beds and seedlings are transplanted when these get adequately infected.

Layering or padding: Inoculum is placed in layers or pads beneath the seeds such that the roots penetrating the inoculum will get infected. Presently this is one of the most effective methods of inoculating plants with VAM fungi. Layering on small and large scale can be done by using manual planters and tractor drum seeders, respectively.

Recommended Biofertilizer application in Moringa (Ponnuswami, 2012)

1. **Seed treatment:** Treat seed with *Azospirillum* @ 200 g /kg of seed.
2. **Nursery:** Apply *Azospirillum* to soil @ 5 g/m^2, phosphate bacteria. 5 g / m^2 and VAM fungi @ 60 g / m^2.
3. **Mainfield:** Apply *Azospirillum* @ 2 kg/ha, phosphate bacteria @ 2 kg/ha and VAM fungi @ 4 kg/ha to soil.

Fertigation in Moringa

Fertigation is the practice of applying plant nutrients (fertilizer) through irrigation, generally through drip system of irrigation. Fertigation supplies nutrients continuously due to uninterrupted water supply. Almost no research was done on fertigation of moringa. However,

Beaulah et al., (2010) suggested the following fertigation schedule for *Moringa oleifera*.

Table 11.2 Nutrient requirement of moringa for fertigation

Nutrients required per ha: 144:24:48 kg (NPK)
Fertigation dose: 144:6:48 kg (NPK)

Name of the fertilizer	Dose (kg/ha)
Super phosphate	113
19-19-19	32
13-00-45	94
Urea	274

Note: The entire dose of super phosphate should be applied three months after planting as basal, since super phosphate is insoluble in water.

Table 11.3 Fertigation schedule for Moringa

Stage of the crop	Duration of Fertigation	Name of the Fertilizer	Number of times	Dosage (kg) each time
First stage	3 months after planting to 145 days	19-19-19 13-00-45 Urea	11 times 11 times 11 times	2.90 0.70 77.2
Second stage	146 days after planting to 190 days	13-00-45 urea	9 times 9 times	3.6 13.0
Third stage	191-235 clays after planting	13-00-45 Urea	9 times 9 times	5.9 8.8

Integrated Nutrient Management

Generally both organic and inorganic fertilizers (chemical / synthetic) are used in crop production. Normally organic manures are applied as basal dressing before sowing / planting during the last ploughing of land preparation. On the other hand, chemical fertilizers are applied as basal dressings as well as top-dressings. During the second half of 20[th]centuary significant progress in crop production and productivity was achieved world over through increased use of chemical fertilizers, besides high yielding varieties, irrigation and synthetic pesticides. Unfortunately, this approach was not accompanied with any major effort to maintain soil health in terms of its microbiological and mineralogical (nutrients components) aspects which are important for crop production. The use of sub optimal or over doses of nutrients by the farmers has also led to severe depletion of nutrient reserves from the soil resulting in multiple nutrient deficiencies. These factors together suggest a need for reduced consumption of chemical fertilizers and increase the use of organic manures, crop residues and bio-fertilizers in combination with or without chemical fertilizers. This whole approach is described as "Integrated Nutrient Management (INM)".

Integrated nutrient management (INM) includes proper combination of chemical fertilizers, organic manures, crop residues, nitrogen fixing crops and biofertilizers suitable for the system of land use and ecological, social and economic conditions. INM is necessary because unbalanced use of chemical fertilizers could exhibit long term effects on soil fertility and productivity. In intense cultivation, application of more chemical fertilizers (NPK) is not sufficient for sustaining the growth and yield and leads to deficiency in the soil for secondary nutrients and micro-nutrients, which limit the crop productivity. On the other hand, use of organic manures, like green manures and crop residues and biodegradable rural and urban wastes supplement the nutrients, increase the efficiency of nutrients leading to improvement in physical and biochemical and biological properties of the soil. The components of INM are chemical fertilizers, organic manures, green manures, crop residues, biofertilizers and legume intercrops.

Beaulah et al., (2004) investigated the influence of INM on annual moringa cv. PKM-1, encompassing organic manures, biofertilizers and various levels of NPK during February-December period at Coimbatore, Tamil Nadu (India). The results showed positive response to the application of above. Initial vigour of trees was higher with poultry manure (500 g / pit) + neem cake (250 g/pit) + panchakavya (2% spray) along with 150:150:100 g of NPK per tree both in main as well as ratoon crops. Organic and inorganic fertilizers significantly influenced canopy spread, which markedly increased by the application of poultry manure + neem cake + panchakavya to 1.87 metres and 1.88 metres at harvest stage in main and ratoon crops respectively. Increased dose of fertilizers also favourably influenced canopy spread at harvest. The combination of poultry manure + neem cake + panchakavya + increased doses of NPK has emerged as the best treatment since it recorded the highest canopy spread of 1.93 metres in main crop and 2.04 metres in the ratoon crop at harvest. The increase in canopy spread was attributed to better synthesis of assimilates in the large photosynthetic area provided by number of branches, more number of leaves and optimum supply of nourishment to meristematic tissues which in turn was due to nitrogen being associated with the synthesis of amino acids might have increased the meristematic activities at higher dose leading to better growth (Beaulah et al., 2004).

Rajeswari and Kader (2004) carried an investigation to determine the effect of INM on growth, flowering and yield of PKM-1 moringa. The investigation clearly brought out that 56.25, 22.5, 45 g NPK + 10 kg FYM + *Azosprillum* (2 kg/ha) + phosphobacteria (2 kg/ha) influenced most of the growth, flowering and yield attributes. This treatment, especially showed a marked influence on number of fruits per plant and there by resulting in the highest yield per plant. It also showed a marked effect on the number of fruits per inflorescence, fruit set per cent, fruit length and girth, fruit weight and number of fruits per plant, which resulted in highest yield of fruits (about 46 t/ha). The yield increase was attributed to synergestic effect of biofertilizers with organic and inorganic nutrient sources brought about additional nitrogen fixation and production of phytohormones and their favourable effect on plant growth by *Azospirillum* and phosphate solubilizing bacteria and FYM. Thus, per plant application of 56.25, 22.5: 45 g of NPK + 10 kg FYM + *Azospirillum* + Phospho bacteria could be adjudged as the best treatment combination to maximize the yield of moringa cv. PKM-1, hence recommended.

Application of FYM (10 t/ha) + 50% chemical fertilizers (50:25:25 kg NPK/ha) + biofertilizers (*Azospirillum* + phosphate solubilizing bacteria) recorded significantly maximum growth (plant height, number of branches, stem girth, canopy spread) and yield attributes (number of pods, pod length and girth and pod yield per plant) (Kadam et al., 2006).

Prabhakar and Hebbar (2007) reported that the yield and yield components significantly increased with application of 15 kg FYM per tree, with 5 kg / ha of bio-fertilizer (*Azospirillum*) and phosphate solubilizing bacteria as compared to other organic treatments.

Trials on INM studies revealed that use of poultry manure (500 g) + neem cake (250 g) + 150:75:75 kg of urea, super phosphate and muriate of potash / plant in two split applications was superior in terms of higher yield, quality (carotene content of pods) and higher nutrient uptake by the plants (Madalageri et al., 2010).

A judicious combination of organic and inorganic fertilizers can sustain commercial production of moringa than the exclusive use of either organic manures or inorganic fertilizer (Beaulah et al., 2004). The concept of INM towards better crop production has emerged as a proven technology. However, not much research work was done on INM in moringa (Rajeswari and Kader, 2004).

CHAPTER 12

Irrigation

Drumstick (*Moringa oleifera* Lam) is a tropical drought tolerant plant and exhibits certain degree of xerophytic behaviour and with stands drought to some extent and does not require much irrigation for its growth and production (Nigam and Mishra, 2003). Under regular and well distributed rainfall conditions moringa can be grown without irrigation. In fact, in the past when grown in home steads, field bunds, group of two or three at cattle sheds, waste lands and before the start of regular cultivation moringa was grown as unirrigated or as rainfed crop and trees yielded well also. But with the advent of commercial cultivation the practice of irrigating the crop for higher production was started. In very arid (dry) conditions and under intensive cultivation, moringa needs irrigation. If the rainfall is less than 800 mm, irrigation facilities have to be provided for production. The need for irrigating moringa under commercial cultivation indeed increased much. Irrigation needs, frequency of irrigation (interval) and quantum of irrigation differ from type of moringa, stage of the crop growth and development and soil type.

Perennial types of drumstick require comparatively less water and irrigation because they are featured with large size water storage cells in limbs and low transpiration rate to tolerate drought condition and also have the adaptation behavior of shedding leaves to further reduce transpiration loss. In contrast annual types require more water, frequent and regular irrigation, because in these types, the roots are succulent and fast sprouting and draw more water from the soil. Annuals need irrigation at an interval of 7-10 days, whereas perennials, being hardy can be irrigated at an interval of 10-15 days based on the soil type (Ponnuswami, 2012).

In case of nurseries, watering / irrigation should be given prior to seed sowing and on the day of sowing. Water stress at this stage causes poor seed germination and subsequently poor crop stand. However, seeds can germinate and grow without irrigation if sown during rainy season. The roots develop in about 20-22 days and allow young plants to endure droughts (Fugli and Sreeja, 2011). After germination the sprouted seedlings may be irrigated on alternate days till they attain a height of 18"-20" and thereafter at weekly intervals till they are lifted for planting in the main field. Care must be taken to avoid moisture stress and wetness in the nurseries till the germination of seeds. Further, there should be no water stagnation around the roots in the nursery, lest there will be root rot. Before lifting the seedlings for transplanting, nurseries may be lightly irrigated / watered for easy lifting.

Similarly in case of limb cuttings, watering should be done before and after setting the cuttings in the field. Thereafter weekly irrigations are necessary to encourage rooting of cuttings. Once the cuttings are established well monthly irrigations would be adequate (Verma and Chaarasia, 1993).

Pits should be watered before planting and after planting of seedlings and three days later to promote root development for survival and establishment. Any water stress at this time is detrimental for survival. Thereafter, the newly planted seedlings may be irrigated regularly during the first two-three months depending upon soil type, receipt of rains and other weather conditions for fast and optimum growing of seedlings. Once established, the seedlings require few irrigations depending on the type of soil and other weather conditions or may be rarely irrigated. Since they pass through monsoon rains, the well rooted cuttings tolerate drought and need irrigation only when persistent wilting of cuttings / trees is evident (Palada and Chang, 2003).

Growing perennial type moringa plants may not require irrigation except during hot weather (Muthuswamy, 1954). In case of annual moringa a shorter period of 10-15 days water stress immediately after pinching is useful for the production of more number of lateral branches and roots.

Frequency (interval) of irrigation depends on climatic conditions, type of soil, stage of crop growth and other crop management practices. In rainy season based on soil type and need a 10-15 days interval and in summer 5-7 days interval may be followed for irrigating moringa. In light soil, irrigation may be given at an interval of 10-15 days, whereas in heavy type soil being water retentive an interval of 15-20 days may be followed. The following frequency of irrigation was suggested for moringa in South India, particularly for Tamil Nadu by Rajangam et al., (2001):

Annual Moringa

1. First two months once in a week (7 days)
2. Next 2-4 months once in two weeks (15 days)
3. Next 4-6 months once in 3 weeks (25 days)

Thereafter, irrigations may be scheduled depending on the time and season of flowering, fruit growth and maturity of pods.

Generally, drumsticks in India are irrigated through basin and channel system of irrigation. However, in areas of scarce available water and dry areas, drumstick can also be grown under drip irrigation system advantageously. Under drip irrigation system trees will be precocious, regular in fruiting and produce higher yields (Singh, 2011). Further, drip system of irrigation provides higher water use efficiency (WUE) (30.58 kg/m^2/ha) (Raja Krishnamurthy et al., 1994). With the same amount of water of basin system of irrigation, four times of area can be covered by drip system of irrigation and yield can be tribled. Additionally, it is possible to take 4-5 ratoons under drip system of irrigation. Annual types responded well to drip irrigation and could give yield of two fold (57%) as compared to rainfed moringa crop (Raja Krishnamurthy et al., 1994) and almost 50% more than conventional system of irrigation. In view of the above, moringa farmers are advised to adopt drip system of irrigation particularly in drought prone areas with profitability.

Soil moisture management in relation to flowering and fruiting

Soil moisture content, rainfall and irrigation have close relationship / role with flowering and fruiting especially in case of flowering in drumstick. Like fruit crops, moringa needs water stress prior to flowering for blooming. During water stress period vegetative growth is suspended/checked to save and provide reserve food for flowering and subsequent production. In arid climate, moringa induces flowering when irrigated after a definite period of rest caused by water stress. It is quantified that moisture stress for a period of 20-30 days depending upon soil type is required and beneficial for flower induction in moringa. Irrigation should be stopped after February for plants to take rest, and irrigations during this period (February and March) leads to less flowering and production; on the other hand, in irrigated area of Konkan region of Maharashtra, irrigation should be stopped during October, to provide stress to the plants for more flowering during November-December (Anon, 2012b).

Soil moisture stress for a limited period of time has positive role in flower induction, on flower bud development and flowering, by suppressing the excess vegetative growth. However, water stress should be restricted prior to flowering only and it is not advisable during the period of peak flowering and fruit setting stages. At full bloom irrigation may be reduced, but not stopped, but after fruit set irrigations are required at 10-15 days interval for achieving good yields. Moisture stress during peak flowering and fruit set would be highly detrimental to moringa crop and results in flower drop, pollen dryness, immature fruit drop and desiccation of fruit lets from stalk end to down wards. The vegetative phase of the crop i.e. seedlings to flowering stage (September to February in Tamil Nadu) should have minimal irrigation to prevent the tree from becoming extremely succulent inspite of the fact that period falls into rainy and winter season and water is adequate in the wells. The reproductive phase i.e. flowering to harvest stage (March to July, in Tamil Nadu) should have abundant irrigation at weekly or ten days interval depending on the soil type to sustain the pod growth and development, as the period falls into summer months when the wells are generally dry and there is no possibility of rainfall also. Hence the crop is successful wherever summer irrigation is possible. Hence, regular and frequent irrigation at 7-10 days interval depending on soil type would be necessary to improve the fruit set and fruit development of moringa (Ponnuswami, 2012). However, during peak fruit development and maturation stages moringa needs moderate irrigation, otherwise the fruits become highly succulent, less palatable and have poor post-harvest life. Complete stopping of irrigation is recommended during fruit maturation and fruit colour changes from green to straw colour for seed production purpose (Ponnuswami, 2012).

Cautious irrigations are beneficial in moringa for flowering and fruiting. Moringa should not be exposed to either excessive or under irrigation, both of which are harmful to the crop. There will be flower drop whenever the soil is dry or wet. Similarly fluctuations of irrigations also lead to flower shedding. Water stress causes loss in market values of product (pods) through reduction in yield, fruit size and quality.   Excess watering to moringa encourages excessive vegetative growth at the cost of pod production, loss of valuable water by run off and loss of nutrients by leaching, increases in operation cost for irrigation, reduction in yield and fruit quality and plant diseases like root rot.   Therefore, excess watering and water stagnation in the moringa field should be completely avoided during fruitset, fruit development and fruit maturation stages. Hence, drumstick should be irrigated carefully to maintain optimum content of moisture in the soil.

Further, if conventional irrigation system of basin and channel system of irrigation is followed care should be taken to avoid the contact of water directly with the base of the trunk, since this will weaken the plant and make it vulnerable to winds, whereas if drip system of irrigation is followed the laterals and emitters should be placed at 1.0 to 1.5 feet away from the main trunk or base of the moringa tree to avoid the above possibility.

In water stress sensitive crops such as vegetables, grown for leaves and fruits, farmers should schedule irrigations very carefully to avoid losses from over or under watering (Beaulah et al., 2010). This needs basic understanding of the growth and developmental physiology of moringa and its timely irrigation needs and is very much essential before planning the soil moisture management especially for flowering and fruiting.

CHAPTER 13

Weed Management/Control

Weeds are out of place and unwanted plants that compete with crops for land, light, water and nutrients and sometimes harbour pests and diseases of crops and cause huge losses to crops and farmers. The losses due to weeds include:

1. Direct yield losses of crops resulting from competition
2. Indirect losses from reduced crop quality
3. Increased costs in land preparation and cultural operations and harvesting
4. Harbouring (as alternate and collateral hosts) insect pests and diseases and
5. Removal of large amounts of nutrients from soil

Due to slow germination and initial growth, wider row spacing, slow lateral spread and adequate supply of nutrients and moisture and long duration, moringa suffers from serious weed problems. In the context of adoption of improved agricultural technologies, effective weed management / control becomes even more important, otherwise weeds rather crops benefit from the costly inputs of improved agro-technologies.

Weeds in moringa include annuals, perennials, and parasitic plants. The weed flora differs from place to place due to variation in agro-ecological conditions and management practices. Weeds also appear in both moringa nurseries and field.

Weeds must be removed regularly to avoid competition for nutrients (especially nitrogen), space and water. If not weeded properly, trees produce less foliage and the leaves at the base of the tree turn yellow.

Weed Control in Moringa Nurseries

The site selected for raising moringa nurseries should be thoroughly ploughed and all the emerging / emerged weeds may be destroyed. Additionally pre and post-emergence herbicides may be sprayed on the nursery beds. For control of weeds in nurseries two weedings at 20 and 30 days after seed sowing should be carried depending upon weed intensity and growth.

Weed Control in the Main Field

Due to the reasons explained above, moringa fields are infested heavily with different types of weeds. Generally it is advised to cultivate the field thoroughly before taking up direct seeding / transplanting seedlings to check and suppress early weed growth because after planting it may not be possible for ploughing the field. The field should be kept weed free by regularly cultivating between row spaces (alleys). The tree basins may be cleared of weeds atleast once in a year and light hoeings of the soil under the trees after cessation of rains will be beneficial (Seemanthini, 1964). At any cost the field should be kept weed free for initial two months which is critical period for moringa seedlings for establishment, survival and growth.

Several methods of weed control were detailed by Ponnuswami (2012). Weeds may be removed at critical periods of weed and crop competition for effective control of weeds. In subtropical climate of India the critical period was observed during 30-90 days after sowing. Weeds may be controlled in moringa by mechanical and physical methods, which are relatively easy, cheap and do not leave chemical residues in the soil, but damage the soil structure and promote soil erosion. Shallow rooted weeds can be controlled by hoeing with hand tools or with intercultural operations. Generally, 3-4 hoeings may be required particularly after each irrigation to avoid crusting of the soil surface during vegetative phase of moringa. But mechanical weeding is laborious and expensive.

Many weeds can be destroyed by simply pulling them out of soil. This is best done when the soil is moist and before seed production of weeds. But this is practical only done in case of small patches and tree basins. When the weeds are numerous for hand weeding, the weeds can be effectively destroyed by mowing. This should also be done before seed production and as close to the ground as possible.

Cultural practices like crop rotation, tillage, intercropping, cover cropping, mulching are also useful in controlling weeds. Certain weeds have association with moringa and mono-cropping and continuous cropping of moringa on the same field increase the intensity of weeds. In such cases moringa may be rotated with other crops. In case of annual moringa, crop rotation is possible after three years of crop. Tillage and cultivation are the traditional means of weed control in agriculture. Inter-row cultivation is another method of weed control and is best done after precipitation.This avoids compaction of soil, breaks surface crustings and destroys weeds on their germination, the most vulnerable stage.

Intercropping is often practiced in moringa as the crop is planted at wider spacing. In intercropping selection of right intercrop is very important. Short duration and quick growing intercrops and moringa amicable and unharmful crops need to be grown as intercrops. Also they should not be collateral hosts of moringa insect pests and diseases. Sunflower as intercrop is particularly recommended for helping to control weed growth in moringa (see intercropping).

Cover cropping is also helpful in checking weeds. The cover crops, which show allelopathic effects on weeds include sunflower, sorghum and rapeseed. Sweet potatoes inhibit the growth of yellow nut sudge, velvet leaf and pig weed. Besides cover crops, their residues left in the field also suppress weeds.

Mulching is an age-old practice especially in horticultural crops, worldwide. The primary purpose of mulching is to check moisture loss from the soil by evaporation. However, mulches are useful for checking weed growth also. Under mulching the seeds of weeds fail to germinate and germinated seedlings will be killed. Milching the tree basins below the tree canopy will also suppress weeds (Palada and Chang, 2003). Mulches include natural ones like different types of straws, crop residues, saw dust etc and synthetic ones like polythene sheets. Among the synthetic mulches black polythene sheet was found more effective in suppressing weeds. Low density polythene sheets of 100 guage thickness will check weed growth by arresting photosynthesis of weeds and ultimately killing them. Further, black polythene mulch increases soil temperature as well as conserves the soil moisture by preventing the evaporation of water from the soil surface.

Chemical Weed Control

Of late chemical weed control with herbicides (weedicides) has been attempted in moringa. These herbicides are sprayed on the growing weeds in the inter row spaces in the field. Use of herbicides however, has not been standardized in moringa yet. However, as a prophylactic measure, a non-selective herbicide viz., glyphosate (round up or kleen up) was suggested before field preparation to control emerged weeds. Unfortunately, however, this herbicide has been banned for use. In view of the above, further research should be carried to find out another safe herbicide (to humans and animals) for management or control of the weeds in moringa.

Biological Control

Living organisms, particularly microbes can be used as weed killers. However, no head way was made in this method of weed control in drumstick.

Integrated Weed Management

Integrated weed management (IWM) is a technology using all the above weed management methods. A long term weed management plan that considers all available management / control techniques to control weeds can be developed area wise for moringa. This long term approach of IWM should reduce the extent of weeds as well as weed seed stock in the soil, without degrading the land and its ecology and agricultural crops.

As farmers are showing interest in the cultivation of moringa, particularly annual types, it is the need of the hour to initiate and intensify research work on weed management in annual moringa to work out IWM strategy considering all the above advantages and to increase the overall profit of the farmers.

CHAPTER 14

Inter Cropping

The practice of growing of short duration crops in the inter row spaces of long duration crops (perennials) is termed as "Inter cropping". It is an agricultural cropping system in which amicable annuals, biennials and short living crops as intercrops are grown in between the rows (alleys) of perennials during their pre-bearing period at the same time and on the same land with an idea of rationally utilizing empty interspace and getting additional income from the intercrops. Intercropping of row spaces is an important and remunerative cultural practice in perennial crops. Intercrops share the same resources along with the main crop (i.e. perennial crop), but give additional income to the farmers. Moringa is basically a perennial crop. Though moringa is spaced less (2 to 5 metres on either side) it offers excellent chance for raising intercrops profitably. In general drumstick trees were grown under monoculture system as pure plantation in the past. However, it is advised to grow intercrops in moringa, which is practiced currently now.

Moringa is a drought hardy plant, hence requires less water for growth and production. In general sandy loams are suitable for moringa cultivation. For intercropping in moringa field the inter crops should also be drought tolerant with same soil requirement; water requirement of both main crop and intercrop should be the same and are the key factors in determining the combination of both crops in intercropping systems. Often climbers such as different types of gourds, pole beans etc., are grown as intercrops in mature moringa, which act as supports for climbing since the vine growth can choke off the young trees. Root system of moringa trees is deep hence does not compete with intercrops for nutrients and water. Shade provided by the moringa trees will be beneficial to those crops which are less tolerant to direct sun light. In other words, one should chose crops which are less tolerant to sun light and tolerate partial shade. The relatively open canopy of moringa tree lends itself well to intercropping in moringa plantation (Beaulah et al., 2010). Moringa sheds a lot of foliage.

However, the leaf litter of moringa has the potential to enhance the fertility of the soil and improve the size and quality of vegetable intercrops.

However, during the first four months after the emergence of seedlings in direct seeding and transplanting in the field, intercrops should not be grown; but can be grown preferably from the second year onwards and continued till the moringa plants become dense and cover the interspace. During the initial period of growth synergistic intercrops may be grown (Beaulah et al., 2010).

Several vegetable crops (okra, tomato, brinjal, onion, cucurbits, radish, carrot, French beans, garden beans, cluster beans, cabbage, chillies) and field crops (maize, sweet corn, mungbean, sesame, blackgram, cowpea, groundnut, *Aloe vera*, tobacco) were suggested as intercrops in moringa. However, one has to be choosy in selecting the correct crop (s) as intercrop (s) in moringa. Crops that tend to shade moringa, should not be chosen, as they will reduce the growth of moringa. Select crops that are adapted to alley cropping such as shade tolerant leafy vegetables, legumes and herbs. Good example would be moringa + cowpea, moringa + cabbage association. However, moringa is highly competitive with brinjal and sweet corn as they reduce yields of moringa upto 50%. On the other hand, sunflower is most suitable as intercrop in moringa in view of control of weed growth (Fuglie and Sreeja, 1998).

In arid-regions and under rainfed conditions of Gujarat intercropping of drumstick with cluster beans and with cowpea combination was highly beneficial as compared to other intercrops and sole crop. Considering the nature of growth pattern of drumstick, cucurbits (climbers or supportive crops) could be introduced based on multitier crop concept approach (Raja et al., 2006). They reported that the gourds, permitted to climb on moringa yielded higher than ground trailing ones. Highest yield was obtained in ridge gourd followed by bottle gourd climbers on moringa than ground trailing ones. The lowest yield was recorded in bitter gourd in both the methods. Snake gourd, bitter gourd and ridge gourd combination with moringa was not remunerative. Studies conducted at IIHR and in several on farm trials clearly indicated that crops like onion, chillies cluster beans can be successfully grown as intercrops in annual drumstick (IIHR, 2002). In some parts of Tamil Nadu, annual drumstick was cultivated as a border crop and also as mixed crop in chilli and brinjal fields (Kader and

Shanmugavelu, 1982). Green manure crops can also be grown as intercrops to enrich the soil fertility.

Moringa itself can be grown as intercrop in other perennial crops like coconut, rubber, mango and sapota in their initial years upto 5 years. In pure crops of moringa, honeybees can be reared for honey profitably. Additionally honey bees can increase the rate of pollination and indirectly increase the pod yields of moringa by 25-50%.

CHAPTER 15

Cropping Systems

Tree based cropping system holds lot of promising in alleviating the present day problems of agriculture like lower water table, nutrient imbalances, soil degradation, soil water pollution, salinity, environmental pollution and decline in yields and farm profits apart from fulfilling the objectives of economic profitability and sustainability. In the context of the above, to reduce soil erosion, increasing cropping intensity and crop productivity, perennial horticulture crops like drumstick (*Moringa oleifera* Lam), which do not require frequent tillage and intercultural operations, for optimum production can be tried for enhancement of sustainable returns and source conservation in viable cropping systems (Pande et al., 2013).

A cropping system refers to growing a combination of more than two crops on the same space (land) and at the same time sharing the same resources. An ideal cropping system should, i) use natural resources efficiently, ii) provide stable and high returns, iii) not harmful to each other (absence of allelopathy) and iv) ecofriendly (do not damage the environment). The commonly practiced cropping systems include: (a) intercropping systems, (b) alley cropping system, (c) agri-silvi-horticulture systems and (d) agro-forestry systems.

An ideal cropping system should aim to: (i) produce higher yields per unit area through better use of natural resources and external applications if any, (ii) offer greater stability in production under adverse weather conditions and with disease and insect attacks and, (iii) provide equitable distribution of farm resources.

In recent years several integrated cropping systems have come into vogue in moringa in South India, as follows.

Intercropping: (See intercropping of moringa). This is the most common and popular system of cropping systems not only in drumstick but also in other perennial horticultural crops such as mango, citrus, coconut, rubber etc. Millets or vegetables are often intercropped in moringa plantations for better use of natural resources like land, water, sunlight etc. and for additional income.

Silvi-pasture System

In this system farmers grow tree species (other than fruit trees) like teak (*Tecona grandis*), neem (*Azardirachta indica*), subabul (*Leucaenalenco cephala*), Casuarina (*Casuarina equistifolia*) along with moringa, mainly on the farm bunds and in some cases as plantation. Moringa has silvi cultural characters such as highest demand of nutrients, provider of little shade and drought as well as frost hardy.

Alley Cropping System

Alley cropping system is the practice of growing food crops in alleys between hedge grows of trees or shrubs which are regularly coppiced or severely pruned. In this system trees are planted in hedgegrows forming wide alleys where in vegetables are planted with the chosen varieties that are adaptable to alley cropping such as shade tolerant leafy vegetables and herbs. With their rapid growth, long tap root with a few lateral roots, minimal shade and large production of high protein biomass, drumstick trees are well suited in alley cropping system. This cropping system is adopted with annual moringa cv. PKM-1 in Tamil Nadu.

Agri-Silvi-Horticulture System

This is a cropping system in which a combination of trees, horticulture crops and agricultural crops are grown on the same land in some form of special mixture or sequence. In this system generally *Casuarina equistifolia* is the tree species (for fodder), *Moringa oleifera* is the horticultural component (for pods) and *Zea mays* (maize) is the agriculture crop (for grain). Agricultural crops such as maize, fodder sorghum, bajra, *ragi*, pulses, oil seeds (groundnut, sesame) etc. can be cultivated with casuarina and drumstick under rainfed conditions. Shanmugavelu et al., (2003) suggested two agri-silvi-horticultural models involving drumstick with casuarina or acacia along with maize for obtaining higher yields and additional income in Tamil Nadu.

Agro-forestry System

Moringa is ideal for agro-forestry as it casts very little shade with deciduous nature, erect growth habit and light foliage and does not affect the agricultural crop in any way (Saroja and Dwivedi 1993). Moringa trees can be sown in the alleys in association with forest trees. In such cases the distance between moringa rows must be 2-4 m and they should be oriented East-West direction to ensure that the intercrops receive adequate sunlight.

Less Significant Cropping Systems

In few pasture locations, extensive leafy types are grown along with natural fence line, which produce predominantly leaves and few (1-2) pods per tree occasionally. Such trees are locally (in Tamil Nadu) are known as "male" trees because of nil/less pod production. Their leaves are used as cattle feed (Ponnuswami, 2012).

Table 15.1 Cropping systems in *Moringa oleifera*

Cropping system	Crops
Inter-cropping system	Vegetables like – tomato, chilli, okra, beans, gourds, leafy vegetables, field crops – sun flower
Alley cropping system	Shade tolerant leafy vegetables and herbs
Agri-silvi-horticultural system	Moringa + casuarina + maize
Agro-forestry system	Agricultural crops: Fodders, legumes, soybean, groundnut, sesame etc., involving drumstick, casuarina + acacia.

Note: It is advisable to avoid associating moringa with crops that require chemical treatment, crops that can compete with moringa, for light (like millets, sorghum), crops that require lot of nitrogen such as maize or cassava. It is better to associate crops that can enrich the soil in minerals, especially in nitrogen like legumes (groundnut, soybean or beans).

Systems of Cultivation in Moringa in Tamil Nadu

The following systems of cultivation are in vogue in Tamil Nadu (Vadivel, 2010).

Home Stead Single Tree System

"Jaffna" a perennial variety of moringa is grown in this system of cultivation in backyards, cattle sheds, and waste lands. The trees are unpruned and allowed to grow as they are. The tree survives on waste water from the house, cattle sheds and rain. It bears only once in a year. The tree is frequently attacked by hairy caterpillar, yet no plant-protection is adopted.

Border Systems of Cultivation

In the dry tract of Moolanaur block of Tamil Nadu, an unique system of cultivation of moringa is being undertaken by the local farmers in which a perennial moringa ecotype (Moolanaur moringa) is planted at the border of every field of cowpea, groundnut, tomato etc., (as main crops). Such main annual crop is husbanded fully and moringa on the bunds shares the inputs and irrigation of main crop partially. This ecotype is extremely drought hardy and sheds entire foliage during extreme drought situation, but revives on the receipt of rains. The pods are available in the rainy season (August – September) when no other moringa pods are available in the market. It is grown as an unpruned plant. The pest and disease incidence is less and no plant protection measures, however, are followed.

Double and Triple Planting

Near Oddanchatran (Tamil Nadu) the farmers prefer to adopt double planting (2 plants / hill) and in some places triple planting is also practiced.

CHAPTER 16

Cropping

Flowering (season)

Moringa (*Moringa oleifera* L.) flowers freely and abundantly. The season of flowering and flowering intensity in drumstick vary from region to region under the influence of rains, temperature, humidity, soil temperature, soil moisture, variety and management of the crop and irrigation (Pugalandhi et al., 2010).

Seed propagated perennial types take 2-3 years for first flowering and fruiting. While vegetatively propagated ones come to flowering within a year. On the other hand, sexually (by seed) propagated annual types come to first flowering and fruiting within 6-8 months of seed sowing depending upon the time of seed sowing (Peter, 1979). Kathiresan et al., (1999) reported that cv. PKM-1 (annual drumstick) comes to flowering 90-100 days after transplanting in the field at Tiruchurapalli, Tamil Nadu.

Moringa trees flower and fruit annually once or twice or more or less continuously throughout the year, depending upon the location with associated weather variables, soil parameters and types of moringa. Further, periods and peaks (months) of flowering also vary between the regions and type of moringa. Ramachandran et al., (1980), reported that in seasonally cool North Indian conditions (i.e. sub-tropical) moringa trees flower only once in a year in summer months (April-June), whereas in South India with hot and somewhat equitable tropical climates trees will flower twice annually. Under consistent seasonal temperature and constant well distributed rainfall flowering can occur twice or even all the year round (Gupta, 2018). For example, in Caribbean islands with seasonal temperature and rainfall regimes, flowering in moringa trees occurred more or less continuously throughout the year (Little and Wadsworth 1964). Mathew and Rajamony (2004) reported that under humid tropical conditions of Kerala, flowering in drumstick was throughout the year except in November and December, but flowering peaked during July-August and March-April. Flowering in both annual cultivars (PKM-1 and PKM-2) also occurred throughout the year with peak from October to November during north east monsoon rainy season and from April to May during summer season (Kanthaswamy, 2005, 2006).

Under Coimbatore (Tamil Nadu) and Bangalore (Karnataka) conditions peak flowering seasons are March-May and July-September. Peak flowering in central parts of Kerala is December-January, while in southern parts it is during February-March and July-August with maximum flowering in February-March.

In general in South India, drumstick flowers in two seasons, though flowers and fruits will be found some times on the trees throughout the years, which is not uncommon, with two principal seasons being July-August and March-April. Twice flowering was also observed in north-western parts of India in annual types and once in the perennial types of moringa (Raja et al., 2008).

Moringa flowers at least four times a year starting from February in China (Jahn, 1996). Flowering occurs during November and December in Sindh and South Punjab, while in central Punjab flowering season starts in January-February in Pakistan (Anwar et al., 2007).

Dry period prior to flowering is necessary for blooming of moringa. However, moringa trees will flower and produce crops (pods) whenever there is adequate water in the soil. Leaf shedding occurs in the months of December-January followed by flower appearance during January to April on new flush of leaves appearing in February-March. Soil moisture stress for a period of 20-30 days depending upon soil type will be beneficial for flower induction in moringa. Soil moisture stress is known to induce flower bud development by suppressing vegetative growth of moringa. Incidentally leaf shedding due to moisture stress reduces transpiration loss of water.

Soil Moisture Management for Flowering

Irrigation has significant role in flowering of moringa. Duration of 10-15 days water stress immediately after pinching is useful for the development of laterals, which show positive role in flower bud development in moringa. On the contrary excess watering and water stagnation should be avoided during bud initiation and developmental stages of flowers (Punitha et al., 2019).

How to increase flowering in Moringa?

Enthusiastic moringa farmers frequently ask this question. Flowering in moringa can be increased by timely pruning. Best time to prune moringa for increasing flowering is from mid-February to mid-March (leave branches bearing flowers and pods that exist at this time). As a result of pruning at this period new sprouts will come up and eventually bear flowers and pods. However, avoid peak monsoon periods for pruning. Never prune branch totally, always cut top one fourth of a branch, because the top one fourth area of the shoot is weak and does not store nutrients.

CHAPTER 17

Harvesting

Moringa (*Moringa oleifera* Lam) is grown for leaves, pods and seeds in different countries for their multipurpose uses. In India, moringa is mainly cultivated for pods and to some extent for leaves. In fact the trees are grown for pods are incidentally utilized for leaves also. Whereas in Philippines, Africa etc. it is mainly grown for leaves, both for human consumption and for cattle. Moringa pods are used for human consumption, while seeds are used for propagation, extraction of oil and water purification. Therefore, depending upon the end use of pods, seeds and leaves they are harvested at different stages of their growth and development.

Pods

For Human Consumption

Commencement of first harvest of fruits (pods) varies with type of moringa, methods of propagation and location of cultivation. Perennial types of moringa, raised by limb cuttings take 8-9 months or even one year some times to bear fruit (Pugalendhi et al., 2010). However, they even take 3 years to start bearing when propagated sexually by seed (Joseph, 2007). On the other hand, annual types commonly raised by seeds, sown in July to September start first crop in 6-8 months (180-240 days) after seed sowing (Pugalendhi et al., 2010). In PKM-1 (annual type) first harvest starts 160-170 days after transplanting seedlings in Tamil Nadu (Kathiresan et al., 1999). Though annual types start bearing pods in 6-8 months after planting, regular bearing commences after second year of planting (Muthukrishnan and Seemanthini, 1974). Season of bearing i.e. harvesting season varies from place to place possibly due to prevailing agro-climatic condition of the location and also type of moringa grown (perennial or annual). In case of perennials pods for human consumption are harvested between March and June followed by a second crop in the year during September-October. The period of harvest may extend 2-3 months in annuals (Veeraraghavathatham et al., 1996). Under North Indian conditions, the fruits of moringa ripen in summer (March-April) whereas in South India flowers and fruits appear twice in a year and so two harvests occur once in June to September and next in March-April in a year (Ramachandran et al., 1980). In Chennai (Tamil Nadu) pods come to harvest and available from April to May. In Andhra Pradesh and Telangana States the season of pod harvesting and availability of pods is mainly during April-May to October. However, in some parts of South India the tender pods are available around the year. In India, one can get fresh moringa pods throughout the year because of climate differences in different parts of the country. Fruit harvesting occurs in March-April in Sri Lanka (Burkill, 1966).

Moringa fruits for human consumption (vegetable, pickling etc.) are harvested when they are half mature or immature, while for seed purpose (propagation or oil extraction) fully matured pods are harvested.Pods for human consumption are harvested when they are still young (about 1 cm in diameter) tender, snap easily, before they develop fibre, seed in side becomes hard, green in colour, pliable for eating whole pod like asparagus (Kathiresan et al., 1999), and of sufficient length and girth (Pugalendhi et al., 2010). Pods 50-75 days after flower anthesis are best for vegetable purpose. In early harvested pods, the seeds inside will be soft and edible along with pod. On the other hand, older pods could be fibrous and develop tough exterior (shell) but the pulp and white immature seeds remain edible until shortly before the ripening process begins (Beaulah et al., 2010). Older pods can be boiled and the seeds and white flashy interior (pulp) can be scooped out and consumed. Delayed harvesting of pods beyond March in Himachal Pradesh caused cracking of pods, hence, harvesting should be over by March and should never be delayed (Verma and Chaarasia, 1993).

In general pods require 65-70 days for edible maturity. Fruits took on average 42 days or 65.5 degree days for horticultural maturity, whereas 70 days or 1112.5 degree days for physiological maturity. In summer, fruits attained horticultural maturity faster in 34 days and (634.65 degree days) and in 59 days, the pods attained physiological maturity (1111.68 degree days) at Coimbatore (Rajamony et al., 2004). Immature seeds can be used in recipes similar to green peas.

Time and Method of Harvesting

Generally, drumstick pods are harvested at different times of the day as per the convenience of the farmer and often without stalks (pedicels). However, this time and method of harvest (with or without pedicel) exhibit marked influence on the post-harvest life of pods and also during storage. Sangeetha et al., (2017) investigated the influence of different times and method of harvest on different aspects of post-harvest life of PKM-1 moringa pods and found that harvesting of moringa pods in the morning hours (7 to 9 am) proved better than afternoon and evening harvesting of pods in respect of PLW, retention of colour and firmness of pods, extension of shelf life (6 days) and on changes in chemical constituents (ascorbic acid, crude fibre and protein). Afternoon and evening harvest of pods showed enhanced PLW, enhanced loss of colour, decline in firmness, shortened shelf life, and affected chemical changes in pods during storage at ambient conditions. Similarly fruits harvested with pedicel showed minimum PLW, reduced colour, and firmness of pods, extended the self of life (5 days) and retained more quantity of chemical constituents compared to fruits harvested without pedicel.

In view of the above, it was advised to harvest moringa pods during morning hours (7-9 AM) with pedicel (stalk) intact for better quality and extended shelf life.

For Propagation and Oil Extraction

Fruits mature in about 3 months after flowering and remain on the tree for several months drying and releasing the seeds by splitting longitudinally. Fully matured pods are allowed to dry and turn brown on the tree. Such fruits should be harvested before they split open and spill over the seeds to the ground. In seed farms, the fully mature pods are harvested as early as possible when they turn brown and dry which can split open easily. Ripe fruits are generally available during of May-June months. Harvesting of brown coloured moringa fruits 200 days after anthesis of flowers, results in the recovery of high quality seeds with high germination potential. Pre-mature and green pods contain seeds which would be whitish instead of brown and be hollow and so unfit for seed purpose and maximum oil extraction. Seeds are extracted manually by split opening of pods using gentle pressure on them. Since, they exhibit higher potential for germination, seeds from proximal (near the stalk) and middle portion of the fruit compared to distal end, they should only be collected discarding the seeds from distal end which are small, shriveled and damaged, as they often show poor germination potential (Beaulah et al., 2010; Pugalendhi et al., 2010). Extracted seeds may again be dried in shade for a couple of days and stored suitably until use (see propagation – seed storage)

Harvesting of Leaves

Green leaves are harvested for both human consumption and for fodder purpose from both trees and young seedlings raised for the very purpose of feeding cattle as fodder.

Leaves are harvested when the seedling plant is 45-60 cm or 150-200 cm tall, which will take 3-6 months or a year respectively (Beaulah et al., 2010; Anon, 2015a). For fodder purpose tender shoot tips along with leaves are harvested while only leaves are harvested for human consumption. Early cutting of leaves and shoot tips encourages the plants to put forth new flush of leaves (Beaulah et al., 2010). Young shoot tips are harvested cutting the tender shoots at an angle of 45^0 using sharp tools like shears, sickles, or knives. In case of large scale intensive cultivation fields, mechanical harvesters can be used for cutting (Pugalendhi et al., 2010). If plants are grown at closer spacing and high density, plants are cut close to the ground level to about 10-20cm from the ground surface (Palada and Chang, 2003). Harvesting of young shoot tips promotes the development of side shoots from the stump below the cut. After each cut / harvesting, new shoots and leaves are put forth, allowing material for subsequent harvests. After harvesting of the leaves, new crop will be ready in 45-60 days. Upto nine harvests are possible annually. The left over stump becomes thick and woody after several cuttings, still it continues to put forth new shoots (Anon, 2015a). In Ghana, after the initial harvesting of leaves at 60 days after sowing of seeds, successive harvestings are made at 35 days interval when leaves are richest in nutrients, particularly crude protein content (Amaglo et al., 2007). A total of 7-9 cuttings are obtained annually, depending on the growth of seedlings and care bestowed on the crop. In some production systems, the leaves are harvested every two weeks.

Grown up and bearing trees also yield good quality leaves both for fodder and vegetable purpose. Leaves are harvested at any time of the year once the trees are established and reach a height of six metres. In Gujarat usually the leaves are cut-harvested from trees once in 6 months, because at that time, the leaves are packed with maximum contents of essential nutrients and vitamins (Gupta, 2018). Six metres tall moringa trees are usually cut down completely leaving 15 cm stump from the ground level at an angle of 45⁰ and leaves are harvested. Short stump results in the production of leaves faster and easy to harvest. Keeping the moringa trees low and bushy by regular harvesting produces more foliage and makes it easy to harvest (Ravindra, 2013). Leaf harvesting can also be done by removing and picking them directly off the tree. Leaves are harvested by snaping leaf stem (petiole), which is a quicker way of harvesting leaves from trees, but the disadvantage of this practice is that the subsequent foliage after harvesting would be less vigorous (Pugalendhi et al., 2010). While harvesting leaves from trees, some selected branches may be left for the next harvest and fruiting.

For making leaf sauces, seedlings, tender growing tips or young leaves need to be harvested (Beaulah et al., 2010).Older leaves are more suitable for making dried leaf powder since stems can be removed during the process of stripping leaf lets easily. Older leaves are usually harvested by stripping the tough and wirey stems.

For fresh vegetable purpose, to be marketed, the harvested leaves are bundled and placed under shade to maintain their freshness. Further, moringa leaves can easily loose moisture and leaf lets abscise quickly, after harvesting leaves, therefore should be harvested early in the morning and disposed off by sale on the same day (Palada and Chang, 2003; Beaulah et al., 2010).

Since, moringa branches are brittle and break easily it is not advisable to climb up the trees for harvesting of pods or leaves. Annual moringas bear heavily often leading to breakage of branches due to heavy bearing of pods. To avoid this, the branches may be propped up with suitable supporters (Pugalendhi et al., 2010).

Chemical pesticides are often sprayed on moringa trees and seedlings for the control of different pests and diseases, which remain as residues in leaves and fruits, whose consumption is harmful to both humans and cattle. Hence, it is advisable to harvest leaves and pods before the pesticidal sprays or after a sufficient gap.

Leaves should be harvested at the coolest time of the day, early morning or late in the evening, but there is no dew on the leaves before harvesting, especially in the morning, to avoid rotting during transport.

CHAPTER 18

Yield

Yields of pods, leaves and seeds in drumstick vary widely depending on the location of plantation, season, variety, fertilization and irrigation regimes and incidence of pests and diseases. Moringa yields best under warm, dry conditions with supplemental fertilization and irrigations (Radovich, 2009). Yields also depend on the age of the tree (Verma and Chaarasia, 1993). However, there is much variation in the reported yields of moringa.

Pod Yield

Perennial types of moringa plants established through limb cuttings are generally low producers of about 80-90 fruits per plant in the initial 2 or 3 years. This is attributed to the reason that often there is little or no production of fruits in the initial years. After two years of growth, from third year onwards yields rise to 500 to 600 pods per tree, which is not uncommon in perennials (Joseph 2007; Pugalendhi et al., 2010). A single good tree can yield between 600 and 1600 fruits every year (Ramachandran et al., 1980; Morton, 1991).

Yields also differ with spacing between plants. Pod yield is around 60 kg / tree / year which is equivalent to 24,000 kg/year at 5 × 5 m spacing and 25 kg/tree/ year equivalent to 40,000 kg/ha/year at 2.5 × 2.5 m spacing in perennials (Beaulah et al., 2010). Closer spacing gives higher yield than wider spacing based on population of trees per unit area. A well-developed tree of 5 to 10 years age produces about 10 to 15 quintals of tender pods / tree / year i.e. 1000-1500 pods / tree (Verma and Chaarasia, 1993). Repeated pollarding or pruning of old branches will promote better cropping for a number of years (Muthuswamy, 1954). The economic life perennial drumstick trees is about 15-20 years.

Afterwards yields of pods decline, hence they are felled down at the age of 15 years and fresh plantation is taken up (Beaulah, et al., 2010).

Compared to perennial types, annual types are heavy yielders, precocious, because of which the cultivation of annuals is preferred and area under moringa increased in India as well as in other countries. Each annual moringa tree bears 250-400 fruits depending on the cultivar and tree age (Pugalendhi et al., 2010). Each PKM-2 tree yields 220 pods per year and a full grown mature tree may yield upto 1600 fruits per year in two seasons. In Hawaii, PKM-2 tree yielded 3 or 4 times more pods than local accessions six months after transplanting (Beaulah et al., 2010).

Leaf Yield

Leaves of moringa are harvested for human consumption (vegetable), for making leaf products (leaf powder etc.,) and fodder for cattle. The yield of leaves depends on spacing (population) and season, besides management of the crop. At 10x10 cm spacing leaf yield was 7.8 kg/m^2 at first cutting in well irrigated, well drained fertilized beds. Trials in Nicargua with one million plants per ha and 9 cuttings over four years period gave on average production of 580 metric tonnes /ha / year. The yield differs strongly between seasons. The yield was higher in the rainy season (1120 kg/ha/ harvest) than dry season (690 kg/ha /harvest). Obviously lack of adequate moisture and higher temperature during dry periods resulted in lower yields of leaves. Commercial yields during winter months (November to March) showed a decline of 50-100% of summer yields. This reduction in yields was particularly attributed to a function of lower solar radiations during winter months affecting photosynthesis and resultant growth of foliage (Beaulah, et al., 2010). According to Beaulah et al. (2010) moringa produces approximately 600 metric tonnes of leaves and stems / ha/year. With proper planting and management green leaf production exceeds 800 tones / ha / year.

Seed Yield

Seeds are harvested for propagation of moringa, for extraction of oil (ben oil) useful as a lubricant, for making biodiesel and water purifications (See seed production).

CHAPTER 19

Post-harvest Losses

Harvested agricultural produce, particularly perishables are living entities, in the sense they breath and undergo changes during their post-harvest life. These changes sometimes lead to qualitative and quantitative losses which are referred to "post-harvest losses" due to biotic and abiotic influences. The magnitude of post-harvest losses are often substantial and vary with commodity, production areas, season of production and at different levels (field, harvesting, handling, transport, market etc.,). Further, the magnitude of post-harvest loss in perishables like moringa is influenced by a wide range of factors like method of harvest, time of harvest, transport, handling, packing and pests and diseases. The extent of post-harvest losses in moringa in countries like India, is substantially high due to prevailing hot climate in moringa growing areas and lack of improved post-harvest management and technology.

However, no research on causes of post-harvest losses of moringa, their extent, at different levels of marketing etc. had been conducted. However, in a survey of Theni and Dindigul districts of Tamil Nadu an attempt was made to find out the causes of post-harvest losses in moringa by National Horticultural Research and Development Foundation. Post-harvest losses in moringa were caused by both external and internal factors like mechanical injuries, damages due to pests and diseases and physiological deterioration. Losses may occur anywhere from the point, where the produce has been harvested or gathered up to the point of disposal and consumption. Poor handling, unsuitable packaging and improper packaging, during transit were the main causes of post-harvest losses in moringa in these districts.

Table 19.1 Various losses at Post-harvest level in moringa

S.No.	Stage	Loss (%)
1.	Harvesting	4.80
2.	Transit	2.33
3.	Handling (cleaning, grading, weighing and packing)	8.21
	Total	**15.34**

Further, at trader's level the average losses were 3.4%, mainly due to handling and delay in marketing. At the wholesaler's and retailer's level the average losses were 6.89 and 8.63% respectively and the main causes were sorting, grading, weighing and delay in marketing. Survey indicated the losses were more at retailers level as the commodity reaches late after passing through one or two intermediaries.

Prevention of Post-harvest Losses

Post-harvest losses affect either quantity or quality or both. Hence post-harvest losses assume significance both economically and nutritionally. Prevention of post-harvest losses is, therefore, very essential, as prevention amounts to increased production, and increased availability of commodity and thus assumes significance. Curtailing / minimizing the post-harvest losses of already produced fruits is more sustainable than increasing production to compensate their losses and is cheaper than equivalent increases in production (Narayana et al., 1989).

Post-harvest losses can be minimized by growing cultivars having longer shelf life, resistance to biotic and abiotic forces, adopting proper cultural practices in the plantation, use of proper harvesting methods, handling, grading, packing, storage and transportation, control of fruit environment, use of chemicals as pre- and post-harvest treatments and proper marketing procedures.

Reduction in quantitative and qualitative losses can be achieved by harvesting fruits at optimum maturity, ripening (for seed), sorting, grading, and suitable packaging materials and proper packing of fruits, adoption of refrigerated transport and storage under controlled conditions (see storage). Reduction in the period from harvest to consumption / processing by efficient marketing system can considerably reduce the losses. Alternate distribution systems such as direct sale to the ultimate consumer / processor by the producer or cooperatives considerably reduce the losses.

Measures for minimizing post-harvest losses in Moringa

1. Harvest the fruits at appropriate maturity as per their end use. For human consumption pods may be harvested at immature stage, when they are still green and tender without fibre and with soft seeds. Conversely for seed purpose fully matured brown coloured

pods need to be harvested. Harvest the pods for human consumption during morning hours (7-9 AM) with pedicel intact for better quality and extended shelf life. In case of market glut, methods such as delayed or staggered picking, with holding the fruits for a few days (by storage) at the markets may be adopted to avoid / minimize the post-harvest losses.

2. Avoid gunny bags for packing moringa pods, as such measure leads to higher losses. On the other hand, use polyethylene film bags or CFB boxes as they extend shelf life and reduce post-harvest losses (Damodaran et al., 1999).

3. Follow systematic grading coupled with scientific packaging and storage to reduce the post-harvest losses.

4. Post-harvest losses during transit can be minimized by employing fast and economical transport system. Road transport may be preferred.

5. Proper storage of moringa also minimizes the post-harvest losses. Use controlled atmosphere storage at refrigerated conditions to achieve higher shelf-life with minimal post-harvest losses (Selvi and Varadaraju, 2016).

CHAPTER 20

Packaging and Storage of Pods

Production of drumstick pods during peak season often exceeds the local demands resulting in low prices and hence farmers do not get fair prices. It has been estimated that about one-fourth of all moringa produce harvested is spoiled before it reaches the consumer and consumption (Arun et al., 2011) due to their short shelf life of 3-4 days at room temperature. Post-harvest loss is not only in physical parameters (PLW, firmness, colour etc.,) but also in nutritional quality, due to improper post-harvest handling. Pods get damaged during the process of handling, transport and marketing and lack of proper packaging and storage. Therefore, to avoid post-harvest losses and to help farmers to realize fair prices without incurring financial loss even in the peak season and off-season it is essential to adopt proper packaging and storage methods to extend the storage life and availability of pods.

Harvested moringa pods are sent to markets in different types of packages like gunny bags, card board boxes, Cheuthi Sp. fibre boxes (CFB) etc. In Andhra Pradesh and Telangana states harvested pods are packed in gunny bags (Plate 20.1) and disposed to markets. In certain districts of Tamil Nadu, harvested pods are packed in old fertilizer bags and handled with locally available materials and taken to local markets by the farmers for marketing. Each gunny would contain 500 pods and would weigh about 25-30 kg. The whole saler at the market after the purchase of the pods from the farmers, returns the gunnies to the farmer and replaces the produce in his own new gunnies. He removes damaged and over matured pods simultaneously. While packing he does grading the pods according to the length and physical appearance. The fully packed gunny bags ready for transport would weigh about 60 kg and contain 600-1000 pods each. The wholesaler then moves the produce to the terminal markets at far off places by road / rail (Chinnadurai et al., 2010). However, Damodaran (1998) earlier advised to pack pods in CFB boxes of size 80 cm × 30 cm × 20 cm that can hold

5 kg of pods for higher shelf life. In a later study, Damodaran et al., (1999) compared different types of packaging material and found that post-harvest losses in moringa by way of reduction in weight and spoilage were more in gunnies followed by CFB boxes in transit. Packaging in polyethylene bags extended the shelf life, reduced PLW, retained carotene and ascorbic acid contents followed by CFB boxes with coir waste as filling material. PLW was higher in fruits packed in wooden boxes with dried grass as filling materials. In view of these observations, it was concluded that polythene bags of 300 gauge, followed by CFB boxes (Plate 20.2) were better to extend the shelf life of moringa pods with less loss in PLW and quality. In a recent study, Shareef et al., (2019), found that drumsticks packed in HDPE and stored at refrigerated conditions stored better compared to LDPE in influencing wholesomeness and post-harvest storage qualities.

Fig. 20.1 Packing drumsticks in gunny bags.

Fig. 20.2 Packing drumsticks in C.F.B boxes.

Post-harvest treatments with certain chemicals can also extend the shelf life and quality of pods. Giraldo et al., (1977) reported that in many countries of the world, fruits and vegetables are washed in chlorine or potassium permanganate solutions, before packing to reduce microflora from the produce there by spoilage. Davoodi et al., (2007) reported that calcium chloride ($CaCl_2$) maintains the colour of the produce better than other chemicals, reduces weight loss (PLW), retains firmness and extends the shelf life of fruits.

Normally, drumstick pods can store at room temperature for 3-4 days, but this can be extended at the low temperature of 8.0°C upto 20 days (Beaulah et al., 2010). Among post-harvest storage technologies, controlled atmosphere storage (CAS) is one of the promising technologies, which helps in increasing the shelf life of vegetables and fruits without loss in their quality. Selvi and Varadaraju (2016) found that CAS at refrigerated storage conditions could extend shelf life of moringa pods to 3-4 times compared to ambient conditions and the best treatment for extending the shelf life of pods was 14°C plus 4% O_2 and 5% CO_2 in a PVC chamber at Madhurai (Tamil Nadu).

Moringa fruit loses its marketability at ambient conditions within a few days of harvest through desiccation, dehydration and subsequent spoilage and shrinkage. In order to store the fruit after harvest, it is necessary to minimize the rate of desiccation, dehydration, deterioration processes by high humidity storage and post-harvest treatment with eco-friendly chemicals as surface disinfectants. Zero energy cool chamber (ZECC) developed at IARI, New Delhi was used as high humidity chamber in the study, using virosil Agro as the disinfectant. It has been reported that it is an universally applicable disinfectant and is highly effective against pathogenic bacteria, fungi, viruses and it has been approved in Europe and Israel for surface disinfectant purpose at concentration upto 3% (Fallik et al., 1994). Pods of moringa were treated after harvest with virosil Agro (VA) (at 0.2, 0.5, 0.7 and 1.0%) and stored in ZECC under high humidity conditions (RH 90-92%) minimum temperature (15.0°C) and maximum temperature of 20.0°C. On eighth day of storage, the decay of pods was as high as 39.35% but it was significantly low with high doses of VA compared to control indicating the efficacy of VA as surface disinfectant. On eighth day, VA at 0.5% retained maximum ascorbic acid and the treatment with VA 0.7% retained maximum chlorphyll of fruits (0.93 mg/g).

Based on these results it was concluded that eco-friendly chemical VA at concentrations of 0.5 to 0.7% can be used as an effective surface disinfectant to reduce the decay percentage and increase the shelf life of drumstick pods (Naiya and Kabir, 2007).

Application of poultry manure plus panchakavya enhanced the post-harvest life and quality of pods. Treatments with post-harvest products like brine solution (4%), pulp powder and leaf powder (PKM-1) enhanced the shelf life and keeping quality of pods under organic cultivation of moringa (Beaulah, 2002).

The time and method of harvest (with or without stalk) also influence the storage and post-harvest parameters. The combined effect of moringa harvest with pedicel (stalk) had the shelf-life of many days under ambient temperature. Such pods retained their colour and firmness at ambient conditions (Sangeeta et al., 2017) (see harvesting).

Preservation of moringa pods and their quality is very essential due to their medicinal and therapeutic properties (Jamaluddin, 2006). The preserved pods can be made available throughout the year. Arun et al., (2011) attempted to develop a simple and cost-effective method of preservation of moringa pods, besides keeping a check on microbial growth. They tested different methods of drying pods. They observed that dehydrating the pods by means of oven-drying, wind-drying and solar-drying were equally effective and yielded more or less similar results in combination with germicide salt and turmeric treatments. Dehydration prevented favourable environment for the growth of microbes enabling to store for longer period of time. Therefore, it was concluded that preservation technique involving dehydration can be effectively used to preserve and extend the shelf life of moringa pods without alteration in the nutritional, medicinal and therapeutic properties.

CHAPTER 21

Diseases

Compared to the number and magnitude of incidence of insect pests, the number and incidence of diseases on drumstick are less. No major diseases on drumstick either from India or elsewhere that affect production and economics of the crop seriously are reported. Diseases of moringa are broadly grouped into (a) nursery diseases, (b) leaf spot diseases (c) root and stem diseases and (d) fruit (pod) disease.

Nursery Diseases

Damping off, seedling wilt, and seed borne diseases are included in nursery diseases.

Damping Off

Damping off is a soil borne fungal disease caused by several fungi. It has been reported in India, (Pande and Ghate, 1998; Beaulah et al., 2010), Cuba and Nicaragua (Lezcano et al., 2014) (Plate 21.1 and 21.2.).

Fig. 21.1 Damping off of seedlings in nursery beds.

Fig. 21.2 Damping off of seedlings ion seed trays.

Symptoms and Damage

Poor seed germination and early death of seedlings in the nursery are attributed to seed borne diseases (see later). Very high seedling mortality (25-75%) was observed due to damping off. In pre-emergence damping off seedlings disintegrated before they come out of soil, leading to poor seed germination. In post-emergence damping off the disease is characterized by development of disease after seeds have germinated and emerged out of soil, but before the stems are lignified. Water-soaked lesion formation occurs at the collar region of seedlings. Infected plants shrivel and collapse as a result of softening of tissues of stems. Disease appears in patches both in nurseries and field. Progressive reduction in seed germination, seedling growth, biomass production and survival with increased level of infection occur (Harsh and Ojha, 2002).

In *Rhizoctonia solani* attack infected stem becomes hard, but thin (wiry) and infected seedlings topple. On the other hand, *Colletotrichum demeticum* (Pers) Grove was mainly linked to such symptoms as chlorosis, stem necrosis, and leaf spots small and rounded light brown at the centre and dark brown at the edge, located on both sides of leaves and spread over the leaf lamina, while *Fusarium solani* (Mart) Sacc was related to stem spots and necrosis, wilting and death of *Moringa oleifera* seedlings (Lezcano et al., 2014).

Etiology

The following fungi were reported in association with moringa seeds in Egypt (Riad et al., 2014): *Rhizoctonia* spp., *Rhizopus* spp., *Trichoderma* spp., *Fusarium solani.*, *F.oxysporum*, *Macrophomina* spp., *Helmintho-sporium* spp., *Penicillium* spp., *Aspergillus niger* and *A.flavous*. The presumptive diagnosis indicated the presence of *Colletotichum demeticum* (Pers) Grove and *Fusarium solani* (Mart) Sacc in association with the disease (Lezcano et al., 2014). Harsh and Ojha (2002) recorded *F.accuminata* (*Gibberella accuminata*) with the disease. According to Satyagopal et al., (2014) *Rhizoctonia solani* attacks the moringa seedlings in nurseries in India. Seeds that failed to germinate showed the association of *F.solani*, *F.oxysporum* and *Rhizoctonia solani*, *S.rolisii* and *M.phaseolina* while germinated and sterilized seeds showed no incidence of fungal attack (Riad et al., 2014). According to Satyagopal et al., (2014) damping off in drumstick nurseries is caused by *Phythium aphanidermatum* (Edson) Fitzo, *P. debaryanum, R.hesse* and *Rhizoctonia solani*, J.G. Kuhn.

Epidomiology

Primary resource of fungi is oospores in soil in case of *Pythium, Sclerotia*, in case of *Rhizoctonia* on the other hand, secondary sources of the disease are zoospores spread through irrigation water, in case of *Pythium*, mycelial growth in soil and Sclerotia through irrigation water in case of *Rhizoctonia*.

Favorable conditions for infection vary from fungus to fungus. For *Pythium*, heavy rainfall, excessive and frequent irrigations, poorly drained soils, close planting, high soil temperature with temperature (around 25-30⁰C). High temperature around 30-35⁰C coupled with high soil moisture favour the incidence of *Rhizoctonia*.

Control and Management

Since the disease is seed borne it was suggested to sterilize the seeds with hypochlorite or another material or procedure to ensure good seedling emergence and control of seed and soil borne fungi. Sterilization of moringa seeds as well as planting soil prior to sowing significantly reduces damping off on moringa seedlings. Damping off of annual moringa seedlings can be controlled by drenching the soil with copper oxy chloride (Beaulah et al., 2010).

Cultural control of damping off includes, raising of nursery in light soils with proper drainage, burning farm trash (waste) on the surface of the beds (for sterilization of soil), sowing seeds on rised beds of 6"-8" height (15 cms) and use of optimum seed rate (600 g/ 40 cm^2). Seeds may be treated with *Trichoderma* and *Pseudomonas fluorescence* and soil application of commercial pesticides are also useful for the management of the disease.

Fusarium (Seedling) wilt

Seedling wilt appears in nurseries during rainy season, causes severe defoliation and wilting of seedlings. It is reported from Konkan region of Maharashtra, India (Fugro et al., 1996).

Symptoms and Damage

The affected plants exhibit yellowing of lower leaves in initial stages, followed by discolouration soon. The leaves wilt, dry and drop. The disease affects a few branches in a plant or the entire plant may wilt irreversibly. The affected plants or branches dry up. The vascular bundles become brown. Plants are usually stunted in growth.

Etiology

The *Fusarium* wilt was reported to be caused by *Fusarium pallidoraseum* Sac by Mandokhot et al., (1994). However, recently Ponnuswami (2012) reported *F.oxysporum f* spp. as the causal pathogen of the wilt. The mycelium is septate and produces macro and micro condia and chlamydospores.

Epidemiology

The disease development is favoured by alternating high and low soil temperature and high humidity levels. Other favourable factors include light sandy soils, low soil moisture level and lower pH. The presence of nematodes in the rhizosphere of moringa greatly increases the infection (Ponnuswami, 2012).

The disease is soil borne and survives in the soil as chlamydospores or as saprophytically growing mycelium in infected debris of crops.

Wind borne spores, surface drainage water and agricultural implements also aid the spread of the disease from field to field.

Control and Management

Soak the seeds in carbendazim (2 g /kg) for 24 hrs prior to sowing. It reduces the incidence of the disease. Drench the nursery soil with copper oxy chloride (2 g/l) to prevent the wilting of seedlings (Nigam and Mishra, 2003). Plant the seeds in disease and nematode free soils. For effective control of the disease, five sprays of carbendazim may be given at 15 days interval starting from the germination of seeds. Carbendazim sprays significantly recorded the lowest (1.05%) incidence as against 94.36% without carbendazim treatment (Fugro et al., 1996). The infected plants should be removed from the site and destroyed. Long crop rotation with non-host crops helps in reduction of the mycelium and source of infection (Ponnuswami, 2012).

Seed and Soil Borne Disease

In general nursery diseases are due to seed and soil borne pathogenic fungi. They infect seeds and roots and become a serious constraint to nursery production as they affect seed germination, seedling emergence and delayed development of seedlings. Emergence and initial growth rate are influenced by soil and seed borne pathogenic microflora, which cause seed rots and seedling growth after emergence. Damages such as seed deaths, seedlings and plant deformities or decreased seedling vigour are caused by seed borne disease. Seed and soil borne fungal plant pathogens cause disease of roots and stems, thus disrupting the uptake and transportation of water and nutrients from the soil. This may result in the appearance of symptoms similar to drought and nutrient deficiencies, which include yellowing, stunted growth and plant death (Riad et al., 2014).

To overcome the seed and soil borne diseases, sterilization of seed and soil with sodium hypochlorite or any other material or procedure prior to sowing of seeds should be done to ensure better and higher germination of seeds and good seedling emergence and control of seed and soil borne pathogenic fungi (see diseases/propagation). Sterilization of seeds reduces the number of contaminated fungi associated with moringa seed from 90% in case of non-sterilized seeds to about 60% with sterilized ones, also decreases the contaminated seed and infected seed by fungi from 80% compared to 19% of non-sterilized seeds. Sterilization of seeds as well as soil prior to sowing markedly reduces damping off disease both at pre-and post-emergence damping off on moringa seedlings, as compared with non-sterilized seeds or soil media.

Foot / Root Rot (Diplodia Root Rot)

Foot / root rot is commonly associated with the condition of poor drainage and water logged condition in the field.

Symptoms and Damage

Symptoms of the disease include rotting of bark around the trunk and breaking of plants, rotting of roots and ultimate death of infected plants. Symptoms appear as oozing of brown spores in the form of gum from the affected bark, which remain as an encrustation on the bark. Droplets of gum trickle down from the infected portion of the stem. In the later stages the bark cracks and sheds longitudinally. Leaves turn yellow. Fruit yield is greatly reduced.

Etiology

Root rot is caused by *Diplodia seriata* Denot (*Botryophaeria obtusa* (*Schwien*) *Shoemaker*).

Epidemiology

In water logged conditions, particularly in rainy season, when drainage is poor, roots rot, infected plants show wilt and death of plants (Ramachandran et al., 1980; Fugro et al., 1996). Besides poor drainage and water logging condition, reduced tree vigour, insect damage, poor nutrition, to the trees and old age are the contributing factors of root rot (Ponnuswami, 2012).

The disease survives throughout the year in dead plant parts either on the infected tree or in the soil. The disease spreads through picnidiospores which are disseminated by wind, splashing rain drops, pruning equipment and even animals. (Pugalendhi et al., 2010).

Control and Management

Under very wet conditions, seedlings / cuttings may be planted on rised mounds so that excess water is drained off. Avoid water logging conditions at the base of the tree trunk and in tree basins. Provide facilities for drainage of water and keep tree basins as well as moringa field free of weeds. Keep the trees in vigorous growing conditions. Broken limbs may be cut properly and cut wounds in the bark especially on limbs and forks should be scraped and protected with Bordeaux paste.

Apply *Trichoderma viride* @ 2 kg/acre mixed with 5 kg FYM per tree at the collar region as a biocontrol measure.

Fungicides like potassium bicarbonate or propiconazole effectively manage the disease. Swab the soil around the stem (collar region) with carbendazim (1g/l) or dithane M-45 (3 g/l) or Bordeaux mixture (1%). Spray carbendazim or thiophanate methyl or chlorothalonil for effective control of foot rot.

Powdery Mildew

Powdery mildew on moringa was reported as an alternate host of papaya mildew, which is the main host of the disease (Ullasa and Rawal, 1984). Powdery mildew has wide range of hosts such as chillies, tomato, cotton, onion, weeds like sowthistle and ground cherry (Pugalendhi et al., 2010). The disease appears on moringa, when papaya and moringa are grown side by side. Powdery mildew on moringa was reported from Bangalore as a colletaral host (Ullasa and Rawal, 1984).

Symptoms and Damage

The symptoms appear on the leaves and pods. The symptoms on the upper surface of the leaves appear as small darkened areas which turn to white powdery spots, enlarge and cover the entire leaf lamina. Initial symptoms appear on the ventral surface of leaves as slight yellowing of the leaves either from the margin or towards the mid-rib on the lower surface of the leaves. These yellow lesions have brown necrotic centres and become corky, in advanced stages. Affected leaves curl upwards and are shed profusely. Affected plants lose their vigour and remain barren, with very much reduced pod yield.

Etiology

Powdery mildew is incited by the fungus *Leveillula taurica* (Lev) Salman(=*Odiopsis*) *taurica* (Lev) Salman. It is an obligate parasite having endophytic mycelium. Conidiophores of the fungus are long and multi-branched. Conidia are pyriform and cylindrical, borne singly or in short chains.

Epidemiology

The disease is favoured by warm and high humidity at a temperature range of 24°-26°C. Growth of the fungus is scanty with sparse production of conidia and conidiophores at higher temperature and lower humidity.

Control and Management

Avoid growing moringa adjacent to papaya, particularly when papaya is affected and showing the symptoms of disease. The disease can be managed by timely spraying of wettable sulphur (0.2%) or tridomorphin or miclobutanil at 15 days interval twice (Pugalendhi et al., 2010).

Leaf Spots

Three types of leaf spots caused by different fungal pathogens viz., Cercospora leaf spot, Septoria leaf spot and Alternaria leaf spot are reported on drumstick (Pugalendhi et al., 2010).

1. Cercospora Leaf Spot (Brown Leaf Spot)

Besides moringa, the pathogen has several alternate hosts viz., grain legumes, carrot, brinjal, (egg plant), pepper (chilli), tomato, tobacco, rice, corn, sorghum, oilpalm, cotton, coffee etc (Pugalendhi et al., 2010). *Moringa oleifera* (drumstick) is claimed as the new host of the pathogen (Kumar et al., 2013).

Symptoms and Damage

Symptoms of the disease appear as scattered brown spots on the leaves. The infection spots are amphigenous circular to irregular in shape slightly sunken, spread to cover the entire leaf lamina. Several such spots coalesce and form irregular necrotic areas and lead to blighted appearance of the leaves. The infected leaves turn yellow, and killed and shed. Stomata of the leaves appear as small black dots in concentric rings on diseased leaves (Kumar et al., 2013).

Etiology

The disease is incited by *Cercospora apii* S. Lat. of late, the pathogen of cercospora leaf spot is claimed to be *Cercospora moringae* Thirum and Govindu causing brown spots on *Moringa oleifera* Lam (Kumar et al., 2013). The fungus is characterized by dark brown stromata from which arise fasiculate straight or slightly curved, continuous simple yellowish brown condiophores producing polymorphous cylindrical, fusoid, 1-3 septate, yellow to brown conidia tapering towards the apex (Ponnuswami, 2012).

Epidemiology

Warm temperature, frequent rains and high humidity are favourable for disease incidence and development (Pugalendhi et al., 2010). Sporulation occurs in abundance at 20º-30ºC.

Conidia are disseminated by wind, rain splashes to the leaf surface where they germinate and cause infection. The spots are produced on the leaves on which conidia are formed serving a source of secondary infection. Under unfavourable weather conditions, the fungus survives on the infected plant debris and the cycle is repeated next year (Ponnuswami 2012).

Control and Management

Hot water treatment of seeds, use of disease free seed, proper field sanitation and proper spacing between plants in the field are some of the cultural practices that can reduce the disease incidence. The disease can be controlled by spraying copper hydroxide, carbendazim, mancozeb, combinations, carbendazim – iprodione, azoxystrobin and mycobutanil. Spraying carbendazim, difolatan, mancozeb, maneb zineb, metirana were reported effective for the control of the disease (Ponnuswami, 2012).

2. Septoria Leaf Spot

Moringa oleifera is one of the several hosts of septoria leaf spot like solanaceous vegetables and weeds. Moringa is infected when it is grown near susceptible hosts and susceptible weeds in the moringa field.

Symptoms and Damage

The disease normally appears on the lower surface of the older leaves as small water soaked circular spots. Later the centre of the spots turns to whitish grey or tanned colour with dark brown margins. Such symptoms also appear on stems, and calyx of flowers. Centre of the spots shows minute black glistening pin-head sized pycnidia. In the advanced stages of infection many dark brown pycnidia appear on the centre of lesions. Severe infection causes defoliation during rainy season. (Pugalendhi et al., 2010; Ponnuswami, 2012).

Etiology

The disease is caused by *Septoria lycopersici* (Pugalendhi et al., 2010). The young mycelium is hyaline, thin-walled and sparingly septate. Older mycelium is brown, infrequently branched and septate. In the initial stages of pycnidia formation several hyphae aggregate at certain point and often interwoven to form a mass of hyphal tissues that assumes the appearance of pseudoparenchyma. Pycnidia are sub-globose composed of 2-3 layers of brown-cells. Pycnidia/spores are filiform slightly curved hyaline and septate with pointed or rounded ends (Ponnuswami, 2012).

Epidemiology

The temperature range for infection varies from 15^0 to 27^0C with 25^0C being optimal. A temperature of 20^0-25^0C with 75 to 92% relative humidity is congenial for disease development. The pathogen over winters in the infected plant debris in the form of mycelium or spores or in the debris of solanaceous weeds. The conidia are wind borne, splashed by rains causing infection. Wet and humid weather favours disease incidence and its development. The fungus also over winters on solanaceous weeds like Jimson weed (*Datura stramonium*), horse nettle (*Solanum carolinensis*), ground cherry (*Physalis sub-glabrata*) and black night shade (*Solanum nigram*) (Pugalendhi et al., 2010).

Control and Management

The disease can be managed by avoiding the solanaceous vegetables (tomato, brinjal, chilli, potato) as intercrops, or growing moringa adjacent to such vegetable fields and by keeping the fields as well as tree basins free of above solanceous weeds. It can also be managed by using fungicides like mancozeb and chlorothalonil (Pugalendhi et al., 2010). Spray application of various fungicides such as benomyl, carbendazim, mancozeb, copper oxy chloride, tolpet, metiram, captafol are also effective against the disease (Ponnuswami, 2012).

3. Alternaria Leaf Spot

Alternaria leaf spot causes leaf blight on tomato, potato and pepper (chilli) besides moringa.

Symptoms and Damage

Symptoms appear on leaves as circular to angular, dark-brown spots with concentric rings / circles. The spots coalesce and cause drying and defoliation of trees. Brown/black patches appear on the branches of infected trees (Pugalendhi et al., 2010).

Etiology

Alternaria leaf spot is caused by *Alternaria solani*. Its mycelium is septate, branched, light brown hyphae, which become darker with age. Conidiophores emerge through the leaf stomata from dead centres of the spots. Conidia are beaked, muriform dark and borne singly or in chains of two. Five to ten transverse septa and a few longitudinal septa are present in each conidia, (Ponnuswami, 2012).

Epidemiology

The disease is favoured by high soil moisture, relative humidity, dew and rainfall. The optimum temperature ranges between 28^0 and 30^0C (Ponnuswami, 2012).

Primary infection takes place through conidia on crop debris in the soil. The secondary spread of the disease occurs through conidia developed on primary spots. These conidia are blown by wind, water and insects to neighbouring leaves of moringa or moringa plants (Ponnuswami, 2012).

Control and Management

The on-set of the disease is hard to detect. Once spots have appeared it is often too late to control and defoliation of the infected plant is inevitable. Clearing of weed hosts of the disease should be practiced. Leaves and young shoots should be checked regularly for the symptoms of the disease. Neem seed extract can be sprayed to control the disease. Spraying with fungicide like mancozeb, maneb, zineb, metiram was reported effective for the control of the leaf spots (Ponnuswami, 2012).

Twig Canker

Twig canker is a new disease on moringa reported from Maharashtra, India (Mandokhot et al., 1994).

Symptoms and Damage

Mandokhot et al., (1994) observed lesions with dull green centres surrounded by light brownish margins on twigs, which in due course of time developed into cankers. Such typical cankers were observed on 2-month old seedlings in nursery in Maharashtra which resulted into yellowing of tissues, followed by defoliation, finally leading to death of the seedlings. On the other hand, Satyagopal et al., (2014) observed the first symptom of the disease as clearing of the veinlets and chlorosis of leaves and drooping of petioles. The younger leaves may die in sucessions and the entire plant may wilt and die ultimately. At later stages browning of vascular system occurs, plants become stunted and die.

Etiology and Epidemiology

Twig canker is caused by *Fusarium pallidoroseum* (Cooke) Sacc (Mandoknot et al., 1994). The causal pathogen requires relatively high soil moisture and soil temperature (Satyagopal et al., 2014).

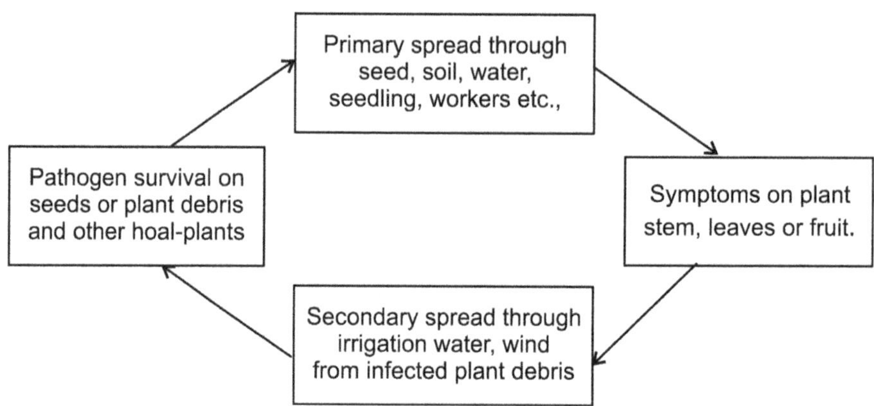

Fig. 21.3 Disease cycle of twig canker (Satyagopal et al., 2014).

Fruit / Pod Rot (Plate 21.4)

This disease was also reported for the first time on moringa from Maharashtra (Kshirsagar and D'souza, 1989) as a new disease and later reported from Bangladesh (Rahman et al., 2001).

Fig. 21.4 Pot rot.

Kshirsagar and D'souza (1989) reported that symptoms of the disease appear all over the surface of the pods, more conspicuously at the stigmatic end., on green pods, elliptical or elongated sunken spots with reddish brown rised margins appear, diseased pods are shrunken to thinning dimension, at their stigmatic ends than healthy pods. In advanced stages of the disease development, the pods are rotten and dried up prematurely, leaving unevenly rised spots over them, pods reaching maturity showed excessive rot. Pod rot is initiated when the pod reaches maturity (Chowdappa and Krishna Kumar 2010). In Bangladesh, the disease appears on immature and also on ripe fruits of moringa and develops rapidly covering the entire fruit. Mature fruits are nutritionally rich food source, because they contain considerably significant amount of proteins and sugars, which attract the disease. Fungal infection affects their nutrition severely and reduce the nutritive value of pods by its consumption (Rahman et al., 2001).

Etiology

More than one fungal species was reported causing pod rot. Kshirsagar and D'souza, (1989) reported *Drechslera hawaiiensis* (Bungi court) Subramand Jain Ex.M.B. Ellis (=*Cochiliobolus hawaiiensis*) Akron = *Bipolaris hawaiiensis* (M.B. Eliss) J.Y. Urchide and Aragalli) after confirmation of its pathogenicity. Earlier, Ramachandran et al., (1980) and later Palada and Chang(2003) also reported this fungal speices as the causal agent of pod rot of drumstick. The fungus was reported on leaves of typha (Ponnappa, 1968), leaves of potato (Khanna and Chandra 1977) and rhizomes (Sharma and Joshi, 1977). The pathogen grows preferably on PDA medium with ashy white to smoke grey, sub-arial septate hyphae sporulating profusely at room temperature (25o-28oC). On the other hand, Rahman et al., (2001) reported that moringa pod rot in Bangladesh was incited by *Rhizopus stolonifer* Ethrenls.

Control and Management

Spraying with chlorothalonil, or iprodione or maneb was effective against pod rot (Ponnuswami, 2012).

CHAPTER 22

Insect Pests and Mites

The cultivation of moringa (*Moringa oleifera* Lam), particularly annual moringa has rapidly expanded and is expanding for the past 2 to 3 decades in India. With this rapid increase there is a corresponding increase in pests and diseases, retarding its potential performance. Productivity and production with the introduction of high yielding varieties, adoption of improved production technology, such as high input agronomic practices, changes in cultural practices, free movement of planting material and changes in weather conditions over the years have contributed to the pest and disease problems, several of which became major constraints in the successful cultivation of drumstick resulting into both qualitative and quantitative losses. New pests and diseases have come into light; once minor pests have become major pests.

Bhutani and Verma (1981) reported 21 different insect pests and two mites from different parts of India on moringa. Later, Honnalingappa (2001) listed 49 insect pests and four mites invading moringa in India. Recently from a fixed plot survey at Bagalkot, Karnataka, Mahesh Math and Kotikal (2014) reported 31 species of insect pests at various stages of drumstick crops. Studies of Brunda Kumari (2014) brought to light 11 species of defoliator pests, of which 4 were caterpillars, 4 species of ash weevils and 3 grass hoppers. Saha et al., (2014) detailed seven insect pests on moringa from Bihar.

The following are some of the insect pests and mites recorded on drumstick in India.

1. Bark eating caterpillar: *Inderbela tetraonis* Moore
2. Hairy caterpillar: *Eupterote mollifera* Walker
3. Moringa leaf caterpillar: *Noorda blitealis* Walker
4. Moringa bud worm: *Noorda moringae* Walker
5. Stem borers: *Inderbela quadrinotata*
    Diaxenopsis apomecyoides Bru
    Batocera rubus Linn.

6. Ash weevils: *Mylocerus* spp.

7. Aphids: *Aphis craccivora* Koch

8. Leaf eating caterpillars: *Teragona siva* Let

                                 Metanastria hyrtacia

9. Tea Mosquito: *Helopeltis antonii* Sign

10. Bud midge: *Stictodiplosis moringae* Mari

11. Scale insects: *Diaspidiotus* spp.

                                 Ceroplatodes cajani Mesk

12. Pod fruitfly: *Gitona destigma* Meigen

13. Thrips: *Scirtothrips* spp.

14. Gram caterpillar: *Heliothis armigera*

15. Looper pests

16. Termites

17. Root and Stem Borer: *Plocoderus ferrugeneus* L.

Mites

1. *Tetranychus neocaledonbus* Andre

2. *Aculus menoni* ChannaBasavanna

3. *A. moringae* ChannaBasavanna

4. *A. pterigospermae* Keifer

The various insect pests attacking moringa trees are generally designated as major and minor pests based on their level of incidence and extent of damage. However, this grouping is not rigid as their status varies from time to time, area to area and susceptibility of moringa varieties. According to the type of feeding Kotikal and Mahesh Math (2016) grouped moringa pests into (a) Borers / internal feeder, (b) defoliators / leaf feeders, (c) sucking insects / sap feeders (d) beetles and weevils. Similarly Joshi et al., (2016) in their review on insect pests of *Moringa olifera* from different moringa growing regions of world, devided insect pests into 4 groups viz., leaf and shoot feeders, sap feeders, flower and fruit feeders and stem and bark feeders. Ponnuswami (2012) categorized the insect pests of drumstick in Tamil Nadu into 5 groups viz., flower feeders, leaf feeders, sap feeders, root feeders and borers.

Four species of insects viz., podfly, leaf caterpillar, bud worm and ash weevils were considered as major pests; while the remaining pests out of 31 species recorded on moringa in Karnataka as minor pests and aphids, white flies and flower thrips as occasional pests were recognized (Mahesh Math and Kotikal, 2014).

The following insect pests of drumstick were recognized as pests of National significance by Agro-Eco-System Analysis (AESA) of Department of Agriculture and Cooperation, Ministry of Agriculture, Government of India, New Delhi.

1. Moringa hairy caterpillar (*Eupterote mollifera* Walk)
2. Moringa bud worm (*Noorda moringae* Walk)
3. Moringa leaf caterpillar (*Noorda blitealis* Walk)
4. Podfly (*Gitona distigma* Meigen)
5. Bark eating caterpillar (*Inderbela tetraonis* Moore)

Root and Stem Borer

(*Plocaederus ferrugeneus* L.)

(Coleoptera: Cerambicidae)

Root and stem borer is a serious pest of moringa trees, especially under neglected conditions.

This pest has eleven other host plants including cashew, silk cotton, besides moringa, of which cashew (*Anacardium occidentale* L.) is the principal host (BasuChaudharyand Mishra, 1977). It was reported from India (Mohapatra, 2006). Africa (Ojiako et al., 2012) and Samoa (Kant and Joshi 2017) on moringa.

Damage

The incidence of root and stem borer is more in mature and bearing trees in neglected gardens. It attacks especially ratoon moringa trees after cutting of the trees. Both young and older plants are attacked. Affected young plants are killed outright, while older ones gradually become weak and die in course of time.

Larva (grub) is the principal culprit. It bores into the trunk / stem just at the place of joint where branches arise and closer to the ground. It makes zig-zag burrows and feeds in side on soft tissues of bark and wood. Presence of white sawdust (chewed up bark and wood) and extrusion of its excreta out of the bored holes on the trunk are the indication of its attack and presence in the trees. After feeding on the above ground stem it bores down in the roots, where full grown grub pupates in a calcareous pupal chamber in between roots and soil.

Due to feeding of this pest, twigs, branches, even whole trees dry up and ultimately the entire tree dies, particularly when more than one grub attacks the host (Pasupathy and Manvel, 2000; Mohapatra and Jena, 2007). Attacked tree can be spotted from a distance due to sparse yellow foliage and drying of twigs.

Bionomics and Lifecycle (Fig. 22.1)

Adult beetle lays eggs singly in the crotches under loose bark. Eggs hatch in 4-6 days with a hatchability of 12.2% in moringa (Mohapatra and Jena, 2007). On hatching, the grub bores into the bark of the tree in its early stages and into wood in its later stages, making extensive tunnels. Grub is whitish in colour with a dark head, measuring up to 7-8 cm when full grown. The grub has powerful mandibles with which it chews the bark and wood. The adult is a dark brown beetle.

Mohapatra and Jena (2007) studied the lifecycle of root and stembore on three host plants viz., cashew, drumstick and silk cotton trees at Bhubaneswar, Odisha (India) and reported that the pest has only one generation in moringa, which is considerably longer in moringa.

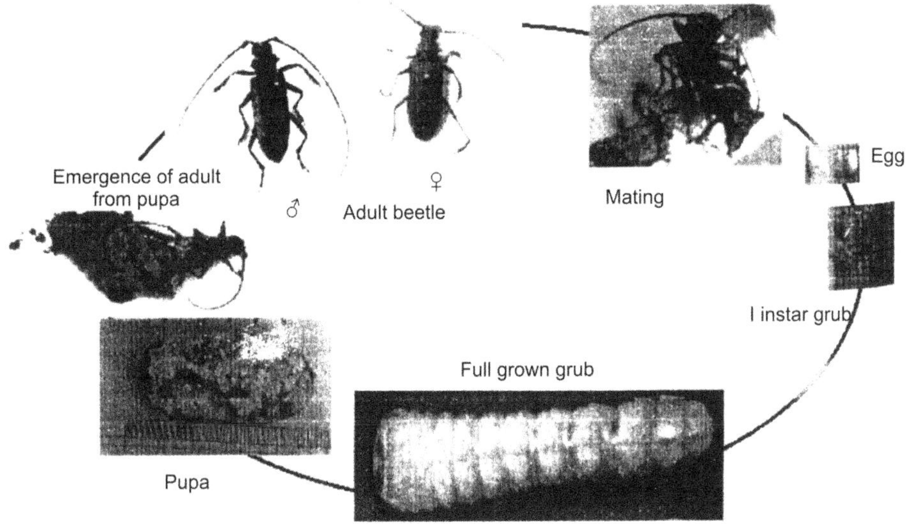

Fig. 22.1 Life cycle of root an d stem borer.

The egg period is 4-7 days, larval (grub) period lasts for 6-7 months and pupal period is 55-60 days. The female adult completes its life cycle in 205 ± 20.6 days on drumstick. The rate of copulation, adult emergence and fecundation per female was less on moringa than on cashew.

Control and Management

The borer escapes the attack of natural enemies and pesticide application because of its concealed conditions in the inter face of bark and inner wood (Mohapatra and Jena, 2007). However, the pest can be managed to a large extent by the following measures:

(i) The moringa farmer should be vigilent for the presence of the pest. Whenever, the pest is noticed with the presence of sawdust (chewed bark and wood) and its excreta on the stem / trunk, scrape the bark with a sharp tool and remove the grub(s) and destroy. In case of deep seated grub(s), kill them by inserting and poking a soft iron spoke (wire) (Lalita Kameswari, 2004). Then insert cotton swab soaked in formalin or celpha tablets and close the holes tightly with wet mud. The grubs die due to fumigation. Alternately syrange dichlorovos (5 ml per hole) or monocrotophos (5 ml per hole) or insert cotton soaked in kerosene or petrol into the hole and plug the holes with wet mud to kill the pest inside (Verma and Chaarasia, 1993).

(ii) Swabbing the scrapped surface of the stem and exposed bark with carbaryl 50 WP (0.1%).

(iii) Painting the trunk with coal-tar + kerosene mixture (1:2) or swabbing with a suspension of carbaryl 50 WP (0.2%) or padding.

(iv) The tree trunks are treated with endosulfan upto a metre length of exposed trunk region after scraping the bark in order to discourage egg laying by adults.

(v) Root feeding with monocrotophos 36 WSC (10 ml) and water (10 ml), kept in a polyethylene bag on the other side (opposite) of the pest attack. The above mixture (20 ml per tree) divided into two equal halves will give protection under moderate incidence condition.

(vi) Swab the trunk with lindane 20 EC (1 ml/l) or carbaryl 50 WP (500 g in 20 litres of water (Pasupathy and Manvel, 2000).

Precaution

In case of incidence of root and stem borer on bearing (fruiting) trees avoid plucking of fruits for 30 days (waiting period) for human consumption or take control measures with pesticides after harvesting the pods (Pasupathy and Manvel, 2000).

Root Feeder – White Grub

Holotrichia reynaudi **Blanhart**

(Coleoptera: Melonlonthidae)

Damage

The grub of the pest feeds on roots, while adult beetle feeds on leaves of moringa.

Bionomics and Lifecycle

Adult female lays eggs in moist soil at 2.5 to 15 cm depth at the rate of 2-4 eggs in a hole. The hatched grub is white in colour and C-shaped. The pest pupates in the soil and undergoes diapause. Pupa continues upto the monsoon rains in June. Adults come out of pupa and soil in June-July. The adult beetle is dark brown in colour measuring about 18-20 mm.

Control and Management

(i)   Deep ploughing of fallow lands and interspaces during March-April should be undertaken with a view to expose immature grubs and pupae for predation by avian predators viz., crows, mynahs, etc. During the ploughing exposed grubs and pupae should be collected and destroyed.

(ii)   Apply well decomposed farm yard manure (FYM) in fields, since the partly decomposed FYM provides congenial conditions for survival of the newly hatched grubs.

(iii) Mechanical collection of adult beetles be done during the nights by jerking the host trees, collecting the fallen beetles and killing them in kerosenized water

(iv)   Cut branches of host trees like neem may be planted in the evening in the moringa plantation to attract the adult beetles for collection and destruction.

(v)   Set up light traps in endemic areas to attract and collect the adult beetles with the onset of monsoon in June-July coinciding with the emergence of adult beetles and the trapped beetles are destroyed (Ponnuswami, 2012).

Long Horn Beetle / Stem Borer (Plate 22.2)

(*Batocera rubus* Linn) (Coleoptera: Cerambycidae)

Long horn beetle is considered to be major pest of moringa. It is widely distributed all over the Indian sub-continent. It is also reported from Asia and Europe (CABI, 2016). Besides *Batocera rubus* others like *Captopsae difactor* Fabricus, *Monohammus* spp. are recorded as stem borers. *Monohammus* spp. is most common in South India.

Damage

Grubs make zig-zag burrows beneath the bark, feed on internal tissues and reach soft wood and cause death of affected branches or stems. Adult feeds on the bark of young twigs and leaf petioles. Grubs seal the entrance of burrows with their excreta, as a result the growing points of twigs and stems get wilted and start drying, shedding of all the leaves (Butani and Verma, 1981).

Bionomics and Life Cycle

Adult female beetle makes small cavities in the stems for egg laying or lays eggs simply in cracks and crevices in the tree bark. On hatching the grubs bore into the stems. Grubs are stout, about 10 cm long, yellowish brown with white spots on elytra and well defined segmentation. Pupation takes place within the tunnels. Adults are medium sized beetles, yellowish brown in colour with white spots on the elytra (Butani and Verma, 1981).

Egg, grub and pupal periods last for 1-2, 24-28 and 12-24 weeks respectively. The pest has only one generation in a year.

Control and Management

(i) Clean the affected portion of the tree by removing all webbed material, excreta etc.

(ii) Next, insert in each hole cotton swab soaked in monocrotophos 36 WSC (5 ml/hole) or any good fumigant like carbon disulphide, carbon tetra chloride, chloroform or even petrol and seal the treated hole with wet mud (Saha et al., 2014).

(iii) To manage the pest, the affected part can be treated with contact and systemic insecticides or fumigants. However, the use of such toxic agro-chemicals is currently not accepted, neither recommended on moringa (Joshi et al., 2016).

Fig. 22.2 Long horn beetle.

Bark Eating Caterpillars

Indarbela spp. (Lepidoptera: Arbelidae)

Two species of *Indarbela* viz., *tetaronis* and *quadrinotata* were reported feeding on moringa in India. These pests are polyphogus in nature, with several hosts in fruit crops, ornamental plants, perennial tree vegetables like moringa, curry leaf and forest trees. These species are well distributed in the entire Indian sub-continent on different hosts including moringa. They were recorded on moringa, where ever it is grown including Bangladesh, Mynmar (Burma), Nepal, Pakistan, Srilanka etc.

Symptoms of Damage

The general symptoms of attack of the pest could be seen by the presence of silken zigzag galleries consisting the bits of bark pieces, chewed material and excreta of larvae. These galleries cover the trunk and branches. When these galleries are removed one can see chewed (eaten) bark and hole (s) in the crotches (forks) of the thick branches with main trunk and within this hole a caterpillar (larva) is found.

Damage

On hatching the caterpillars feed superficially on the bark of the main stem or woody branches making zigzag channels for few days and later bore holes inside the bark in crotches (forks) and remain in the tunnels during day and come out at night and feed on the bark under silken galleries (Usha Rani et al., 2010). The larva eats through the bark into the wooden part under silken galleries. The larva chews out the bark. The larval tunnel remains closed with frass, which is drawn into a sleeve through which the larva moves. The tunnel is used as a shelter by the larva. Severe incidence of the pest may result in the death of the damaged branches / stem (Atwal, 1986) and often damaged branches break either due to high winds or crop load. The incidence of the pest is more in the older and neglected and over crowded garden.

Ali et al., (2007) studied the seasonal occurrence of developmental stages (viz., larva, pupa and adult moths) of *I.quadrinotata* on different growth stages of *Moringa oleifera* in Bihar and recorded high occurrence of the pest during February to October and low to moderate occurrence from seedling to tree stage.

Life Cycle

The adult female lays eggs in clusters (15-25) under the loose bark of the large thick woody branches of host tree. A single female lays as many as 2000 eggs in its lifespan. Eggs are spherical and dirty white in colour, which hatch in 8-10 days.

Freshly hatched larvae nibble the bark wood with their chitinized hook-like pointed structure and after 2-3 days of feeding bore inside the wood in the crotches of branches. Generally only one hole will be observed. But in severe cases, 15-100 holes may be found in a single tree. Larval period lasts for 9-11 months.

With the rise in temperature, the larvae start pupating with in the tunnel made in the tree trunk or main branches. Pupation occurs with cephalic end of the pupa slightly protruding outside. Pupal period lasts for about 3-4 weeks.

The emerged adult moths are short lived. Male moths die soon after mating within 24 hours and the female moths live for 2-3 days more after mating to lay eggs and die after egg laying. There is only one generation per year.

Morphology

Eggs are spherical and dirty white in colour. Larvae are dirty brown, and stout. Full grown caterpillars (50-60 mm long) have pale brown bodies with dark brown heads. The full grown larva is smooth with sparse hairs. The thoracic legs are simple with the last segment ending in a curved claw. The curved end bears several small spine – like processes. These spines and teeth like processes help the pest to orient itself towards the tunnel (bore / hole) mouths prior to eclosion (Saha et al., 2014).

Pupae (18.5 mm long) are stout, reddish, brown in colour with two rows of spines on each abdomenal segment arranged transversely on anterior and posterior margins.

The adult moths are pale brown in colour. Fore and hind wings are creamy white in colour. Fore wings have brown spots and streaks with numerous dark rufous bands of strigae on abdomen and hind wings are fucous. The moth has dark rusty red spots on fore wings in rows. In *I.tetraonis*, there are more intensive spots before the margins and at the margins of the fore wings. The adult is a large sized moth with a wing span of 40 mm in the female and 30 mm in the male of *I.quadrinotata*.

Control

The pest is not seen outside, since it feeds beneath the silken galleries during day time and nocturnal and hides in the bores in the day time. For effective control of the pest, therefore, before under taking any control measure, the webs of galleries on the tree should be removed. Since the incidence occurs in neglected and over-crowded plantation, keep the orchards clean and avoid over crowding of plants.

Physical / Mechanical Control

Kill the larva with in the tunnel by inserting a sharp metallic probe (like a cycle spoke) and seal the entrance with tar or wax.

Biological Control

No parasite and predators have been reported on *Indarbela* species on moringa. However, when the caterpillars fall to the ground they are eaten by ants. Fashih and Srivastava (1988) reported natural incidence of *Beauveria basiana* on the larva, causing 100% mortality within 4-5 days.

Chemical Control

Apply a toxic substance either by injection or by inserting a cotton swab soaked in the toxic substance. Spot application either by brushing or spraying may also be tried in certain cases. Apply 5-10 ml of dilute solution of quinolphos 25 EC at 1:200 or detamethrin 2.8 EC at 1:200 level into bored holes for good control of the pest bores. David and Ramamurthy (2016) suggested injecting chlorpyriphos or profenofos emulsion into the bores, followed by sealing the bore with wet mud. Sharma and Kumar (1986) suggested to take up control measures against the bark eating caterpillars at the nibbling and tunneling stages of the pest.

(i) **At nibbling stage** (i) keep regular watch from July onwards for the incidence of the pest, (ii) as soon as caterpillars start nibbling the bark spray 0.05% endosulfan or 0.05%, trichloride or 0.05%

dichlorvos on the trunks and branches of trees and forking of branches.

(ii) **At tunneling stage:** (i). Fumigate the trees / orchards with fumigants like carbon disulphide, or bisulphide, kerosene oil, petrol, formalin, (ii). Inject cotton swab soaked in a solution of the fumigant (1:100) water into the hole and seal the hole with wet mud.

Termites

(Isoptera: Kalotermitidae)

At times termites may be a problem in moringa especially when limb cuttings are planted in the field. The roots of moringa trees are adapted for water storage and termites find them convenient to attack in search of water. In Puret Rico, moringa is not recommended for planting (Martins and Rubertie, 1975, Parrotta, 2009). Soils that are heavily infested with termites should be avoided for cultivation of moringa, as their control is uneconomical. The following measures are recommended to protect moringa from termites:

(i)   Mulch the tree basins with the leaves of castor plant, mahogamy chips, tephrola leaves or persian lilae leaves

(ii) Heap ashes around the base of the moringa plants

(iii) Spread the dried and crushed powder of stems and leaves of lion-ear or Mexican poppy around the base of the plant.

Thrips

Ramaswamiahiella subnudula **Karmy**

(Thysanoptera: Thripidae)

Ramaswamyiahiella subnudula Karmy a polyphagous pest was found feeding and breeding in inflorescence of drumstick in India (Bhutani and Verma 1981). Thrips are the minor sucking pests on moringa (Plate. 22.3). They suck sap from both new flush and flowers. Feeding on new flush results in stunting of affected portions, turn malformed without flowering. The infested flowers wither and dry up. The damage symptoms resemble those of zinc deficiency (Reddy et al., 2010; Mahesh Math and Kotikal, 2014).

The pest appears throughout the year whenever drumstick is in flush and flowering. Maximum population, however, was observed during the month of September at Bagalkot (Mahesh Math and Kotikal, 2015).

For the control of flower thrips, Reddy et al., (2010) advised to spray either metasystox (2 ml/l) or Acephate (1g/l) during flowering period.

Fig. 22.3 Thrips.

Tea Mosquito Bug

Helopeltis antonii **Sign**

(Hemiptera: Miridae)

Tea mosquito bug is principally a pest on tea; but it also feeds on other plant hosts like cashew, guava, neem, grapevine and drumstick. This is widely distributed in Kerala, Karantaka, Tamil Nadu, Goa and Maharashtra. The pest is serious in Trivendrum and Quilon districts of Kerala, where tea and cashew are extensively cultivated.

Damage and Symptoms

Adults and nymphs of the pest suck the sap from buds, young leaves, tender stems and panicles by puncturing with its needle like stylets and injecting toxic saliva. These punctures appear as reddish brown water soaked spots, which later coalesce together to form necrotic spots. Due to their intensive feeding, leaves curlup, become badly deformed and remain small. Gradually the terminal shoots die, dry up showing die-back symptoms, flowers drop and the whole inflorescence dries up. The surviving fruits show white patches on them, which also dry up ultimately. When young fruits are attacked they fail to develop fully. Attack often is serious and whole plant presents the appearance of a dead tree without leaves and flowers (Pillai et al., 1980).

Bionomics

The tea mosquito bug feeds and breeds on moringa. Adult is a blackmoth with red thorax, black and white abdomen and greenish brown wings. Body is small, slender, with long antenna. An erect knobbed proboscis on the scutellum is characteristic of the species.

Bugs are active in the early morning and again late evening hours (dawn and dusk) and hide in the nearby bushes during the remaining day. White eggs with two filaments arising from the operculum are inserted into tender shoots of the host. The egg period is 5-7 days, nymphal (greenish yellow in colour) period is 10 days with five nymphal instars.

Management / Control

- Collect and destroy the damaged plant parts.
- Spray any of the insecticides: profenofos 50 EC 800-1000 ml, Thiamethoxam 25 WG. 100 g, endosulphan or phosalon or chlorpyriphos or dimethoate at one litre per hectare in 500 litres of water. Spray the insecticide in early morning hours or late in the evening hours on trunks, branches, foliage and panicles for effective control of the pest.

Scales

Caroplastodex cajani **Marshell. (Lab Lab scale)**

Diaspidotus **spp**

(Homoptera: Coccidae)

Scales are minor sucking insects. The lab lab scale (*Caroplastodex cajani* Marshell) was recorded to infest redgram, lab lab beans, ber, tulsi and moringa (Ayyar, 1963), while a scale insect of *Diaspidotus* was noted on drumstick at Madhurai (Tamil Nadu) for the first time during 1958 and 1959 (David, 1961).

The nymphs and adults of lab lab scale occur during January— February and August to December.

Lab lab scale was noted feeding on tender shoots as well as the fruits and stalks. Infested tender shoots eventually dryup. In case of severe infestation, the tender shoots, fruits and their stalks (i.e entire trees) are fully covered by the scales and the tree presents a sickly appearance with few scattered yellow leaves. Eventually the shoots dry and the size of the pods gets affected (Ayyar, 1963; Sivagami and David, 1968). Both nymphs and adults of *Caroplastodex cajani* suck the plant sap and reduce the vigour of plants. Though each scale insect takes only a few drops of sap during its life time, presence of enormous number of scales sucking the sap continuously weakens trees and ultimately affects the vigour of the trees and size of fruits (Usha Rani et al., 2010).

Diaspidotus spp covers the trunk as well as branches giving the tree a sickly appearance with poor fruit setting. The scale insect has a colouration similar to that of branches making it difficult to detect its presence (David, 1961). The adult female is circular and is yellow / light brown about 1.1 mm in length and 1.0 mm in width. It lays light yellow oval shaped eggs. The crawlers are also oval and yellow in colour (David, 1961).

Aphids

Aphis craccivora **Koch (Lab lab aphid / legume aphid)**
Aphis gossypii **(cotton aphid)**
(Homoptera: Aphidae)

Aphids are sucking pests and polyphagous in nature. The above two aphids were reported on moringa (Mahesh Math and Kotikal, 2014). *Aphis cracvivora* Koch was reported from India, Hawaii and Pacific Islands (Morton, 1991; Fuglie and Sreeja, 1998, Palada and Chang, 2003). Host range of *A.craccivora* includes lab lab beans, groundnut, cluster beans etc., and cotton in case of *A.gossypii* (Pugalendhi et al., 2010).

Period of Incidence

Munj et al., (2001) noticed the incidence of aphid, from second fortnight of October till first fortnight of April. It reached peak during the first fortnight of January, which declined gradually during first fortnight of April at Dapoli (Maharashtra). Mahesh Math and Kotikal (2015) observed the incidence of *Aphis craceivora* during second fortnight of February to first fortnight of May at Bagalkot (Karnataka).

Damage (Plate. 22.4)

Both nymphs and adults suck cell sap of leaves, terminal tender shoots and flowers (Sivagami and David, 1968, Munj et al., 2001; Mahesh Math and Kotikal, 2015). Aphids attack the seedlings in the nursery leading to stunting growth and wilting (Rangappa, 1980). Aphids infest tender shoots on the underside of leaves in groups (David, 1958). Aphid feeding causes yellowing and drying of leaves (Mahesh Math and Kotikal, 2014). Feeding of flowers results in shrivelling and drying of flowers and finally shedding of flowers (Munj et al., 2001). Aphids also feed on the terminal end of pods causing tip drying (Honnalingappa, 2001). Cotton aphid causes damage to the tender shoots, while lablab aphid causes damage to leaves (Usha Rani et al., 2010).

Bionomics

Aphids reproduce parthenogenetically, hence their population build up is very fast (Sivagani and David, 1968; Honnalingappa, 2001).

Control and Management

Rangappa (1980) advised malathian (2 ml/l), or monocrotophos (1.5 ml/l) or metasystox (2 ml/l) Imidacloprid (0.25%) or Tatamida (0.25%) in nurseries for the control of aphids on moringa. Population of *A.craccivora* can be effectively controlled by monocrotophos, methyldemeton and dimethoate, which are applied either as foliar sprays at 0.05%, 0.25% and 0.03% respectively or by padding after diluting the formulation. The above insecticides are systemic having acropetal action and pests being found in growing regions, padding method has proved as effective method.

As a precaution all the pods should be harvested before spraying the insecticides.

Biocontrol

For biological control of aphids, encourage the population buildup of *Chelomenes sexmaculata*, the predator of aphids. Release the first instar of *Chrysoperla carnea* @ 10,00,000 per hectare.

Fig. 22.4 Aphids on leaves.

Moringa Leaf Caterpillar

Noorda blitealis Walker

(Lepidoptera: Pyralidae)

Noorda blitealis Walker (leaf webber, leaf eating caterpillar, moringa moth) is categorized as a defoliator of *Moringa oleifera* Lam (Mahesh Math and Kotikal, 2015) and is considered to be the most serious pest of annual moringa as it causes serious damage to the plant. It is an important pest of moringa in South India and South rift valley of Ethiopia (Anjulo, 2009). The pest was first described by Walker in 1859 (Plate 22.5).

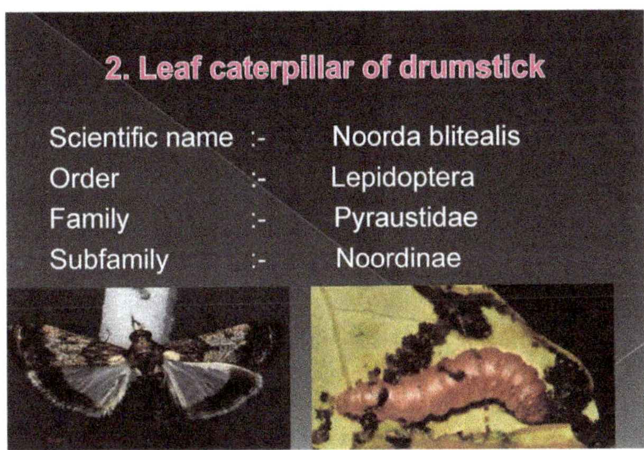

Fig. 22.5 Moringa leaf caterpillar.

As leaf feeder of moringa, *N.blitealis* was reported from India by several workers (Butani and Verma, 1981, Butani and Jotwani, 1984; Nair, 1995; David, 2001). Further, it was reported as a pest of drumstick from many growing countries of Africa. South Africa, Namibia, Gambia, Madagascar, Reunion etc., (Gillett, 1997; Demuelenare, 2001; Parrotta, 2001; Satti et al., 2013), from Nigeria (Ratnadass et al., 2011), Sudan (Satti et al., 2013), Niger (Litsinger, 2014), Burkina Faso (Dao et al., 2015).

Damage

N.blitealis is a serious defoliator of drumstick trees and feeds mainly on leaf lamina. It is a major pest on nurseries and cause 100% defoliation and hence considered a menace (Shamila and Joshi, 1997). In severe attacks 100% defoliation of whole tree occurs, particularly when its control is neglected. Young larvae feed voraciously on the foliage and strip the branches completely (Madhusudan Reddy, 2002). During the severe damaging periods the whole branches *of M.stenopetela* become defoliated and cause shortage of foliage to the rural and urban communities in the semi-arid low land areas of South Africa (Anjulo, 2009). Infested trees become barren with no fruiting and reduced yield.

Larvae make a thin silky web (hence it is often referred to as leaf webber) with leaf lets and tender twigs on ventral side of leaves and feed on them in the web leaving only veins. Early instars of the pest feed on the leaves by scraping the chlorphyll content, resulting in papery appearance of leaves and later instars feed on mature and entire leaves leaving veins behind. In severe infestation the infested plants were almost without leaves resulting in 100% damage to foliage (Munj et al., 1998a; Honnalingappa, 2001).

Besides feeding on foliage, *N.blitealis* also attacks the pods of moringa. After hatching, the caterpillars nibble the surface of the pods, leading to gum exudation of pods. This kind of damage to pods by this pest was recorded for the first time at Varanasi (Halder and Rai, 2014).

Periods of Incidence

Moringa moth occurs throughout the year, however, its period of peak / heavy incidence varies from region to region depending on the availability of green foliage and local weather conditions. In Tamil Nadu heavy incidence of the pest was observed twice, once during March-April and again during December-January (Anjaneyamurthy, 1985). At Dapoli, Maharashtra, the pest was active throughout the year, and reached its peak in the month of January. The pest has shown increases and declines in its incidence. The pest incidence increased gradually from first fortnight of January to first fortnight of August, which declined gradually during the first fortnight of September and then increased in the first fortnight of November, to reach a peak during the first fortnight of January (Munj et al., 2001). Maximum incidence of the pest was noticed in the vegetative stage i.e. from June-July and vegetative grand growth stage (July-August) and minimum damage was observed in flowering stage (September) at Bagalkot, Karnataka (Brunda Kumari, 2014). The incidence of the pest was recorded throughout the year at Bagalkot, but maximum larval population was noticed during second fortnight of April followed by second fortnight of October (Mahesh Math and Kotikal, 2014), while minimum population of the pest was recorded in May, obviously high temperature during May might have drastically affected the population (Mahesh Math and Kotikal 2015). Earlier, Munj et al., (1998) reported the pest infestation in Konkan region having 3 peak periods of defoliations, the first during July to August, second during October and the third during January. The pest was active throughout the year and the maximum population was during January and lowest during May to June.

Bionomics and Life Cycle

Eggs are creamy white in colour and oval in shape and laid in clusters of 34-96, usually on ventral surface of tender leaves. A female moth lays upto 232 eggs in her life time. Larvae are devoid of prothoracic shield. Pupation takes place in soil, adult moths of *N.blitealis* are similar to *N.moringae* but larger in size. Moths of *N.blitealis* are medium sized ones having forewings broader, and wavy with rectangular apex, with erect outer margins and uniform dark in colour with small white streaks at the inner area of the base. Hind wings are hyaline with broad black marginal band towards anal side (Beaulah et al., 2010).

Munj et al., (1998a) observed the life cycle of *N.blitealis* Walk at Dapoli as under.

Table 22.1 Biology (life cycle) of *Noorda blitealis* Walker at Dapoli, Maharashtra (Mean values)

S.No.	Stage of the pest	Mean duration (days)
1.	Egg	2.90
2.	Larva	10.68
3.	Pre-pupa	2.55
4.	Pupa	7.90
5.	Adult-Male	9.10
6.	Adult-Female	15.55
7.	Total life cycle (Egg to adult emergence)	24.03

Under laboratory conditions the incubation period of eggs was 3-4 days, total larval period was about 14 days. Pupal duration was 5-8 days in male and 6-9 days in female. The male and female adults live for 4-9 days and 7-20 days respectively during April to May at Bagalkot. The life cycle was completed in 16-23 days with average of 18.3 days (Brunda Kumari, 2014).

Control and Management

Moringa moth (*Noorda blitealis*) can be controlled and managed by cultural, chemical and biological measures. To manage this pest various practices have been tried such as cultural measures, pesticide sprays, biological control with bio-agents and botanical extracts and search for resistant varieties (Anjulo, 2009; Patel et al., 2010; Satti et al., 2013; Litsinger, 2014; Kumari et al., 2015; David and Ramamurthy, 2016).

Cultural Measures

- Collection (by hand picking) of larvae and destruction of infested leaves with silken webs and caterpillars in the initial stages of infestation.
- Provision of sitting arrangements for birds above the height of moringa crop in the field enabling the birds to sit and prey on the pest (Beaulah et al., 2010).
- Since the pest pupates in the soil, plough around the trees to expose pupae for birds', predators (Satyagopal et al., 2014).
- Arrange light traps @ 1 per ha to attract and destroy adult moths in kerosenized water.

Chemical Control

- As the pest pupates in the soil during February and November, apply phorate granules (10 G) in tree basins and incorporate in the soil to destroy pupae (Madhusudan Reddy, 2002).
- Dust carbaryl at 25 kg /ha or spray carbaryl 50 WP @ 2.0 g/l for controlling budworm, leaf webber and leaf eating caterpillar (Munj et al., 1998b).
- Spray (5 ml/l) quinalphos or monocrotophos (1.5 ml/l) twice at 10 days interval (Madhusudan Reddy, 2002).

- Spray profenofos (2 ml/l) twice at 10 days interval (Reddy et al., 2010).
- Spray malathion (2 ml/l) once or twice to reduce the infestation of the pest.
- Spray dichlorvos (0.04%) and fenthion (0.05%) for effective control of the pest

Bio-control

- Encourage the presence of spiders, the natural predators of the pest to inhibit the infestation of the pest on new flush.
- Biocontrol agents like *Bacillus thuringiensis* and *Aspergillus flavus* formulations may be used for successful management of moringa leaf caterpillar (*N.blitealis*) (Ragumoorthy and Armugam, 1992; Shamila et al., 1996). The pest has shown higher susceptibility to sprays of spore suspension of *A.flavus* with upto 70-80% mortality (Shamila et al., 1996).

Moringa Hairy Caterpillar

Eupterote molliflora (Lepidoptera: Eupterotidae)
Metanastria hyrtaca (Lepidoptera: Lasiocampidae)
Tarasgama siva Lef (Lepidoptera: Lasiocampidae)

Eupterote molliflora (Plate 22.6)

Moringa hairy caterpillar is a destructive and specific pest of drumstick particularly in South India and is common wherever perennial drumstick is grown. This pest has assumed greater significance because of wide spread cultivation of annual drumstick (Parthasarathy et al., 2004). However, according to Reddy et al., (2010) the pest does not attack annual drumstick which is one of the reasons favouring the annual moringa and enhancing the cultivation of annual moringa. It is also assumed that the presence of this hairy caterpillar with its irritating hairs had discouraged the taking up of growing moringa in backyards and cultivation of perennial types of drumstick. In addition to drumstick, it was also recorded on *Ricinus communis, Brassica oleracea and Morus alba* (Saha et al., 2014).

Fig. 22.6 Moringa hairy caterpillar.

Period of Incidence

The incidence of the pest is more during rainy season (north-east monsoon) during October to November in South India. Severe out breaks of hairy caterpillar were reported on moringa in South India (Parthasarathy et al., 2004).

Damage

The larvae of *E.mollifera* feed gregariously by scrapping and gnawing foliage, often resulting in complete defoliation of the tree during severe infestation. It appears right from nursery stage. It is a nocturnal feeder. It feeds during nights and congregates at the base of the trunk of the tree during day time in shade as a protection against sun. Small larvae feed by scraping the stems, while large larvae feed on leaves leaving veins. They love tree gum much.

Bionomics and Biology

Eggs are laid in clusters on leaves and tender twigs. The newly hatched caterpillars are brownish and densely hairy (tufts), which are irritating. Adults are large sized moths with uniform light yellowish brown wings, having faint lines of reddish brown colour spread over the wings. Moths appear with the onset of monsoon rains and lay eggs.

Egg period lasts for 6 days, larval and pupal period lasts for 12-14 and 8-10 weeks respectively. Pupation occurs in soil in earthen cocoons in the tree basin. There is only one generation of the pest in a year.

Control and Management

The pest can be managed by cultural practices and controlled by chemical insecticides.

- As the larvae congregate at the lower portion of the tree trunk, traditionally the larvae were killed by exposing them to fire torches (before the advent of chemical insecticides). The same measure can now be adopted in backyards.
- Collect and destroy egg masses and caterpillars
- Set up light traps @ one per ha to attract and kill the adults immediately after rains.
- Apply neem powder or carbaryl powder or carbofuran granules at the plant base to destroy pupae.

Chemical Control

- Use quinalphos 25 EC (2 ml/l) or carbaryl (3 g/l) sprays changing the chemical every time (Rangappa, 1980) or spray endosulfan 35 EC (2 ml/l) or monocrotophos (1.6 ml) or Dursban (2.5 ml/l) (Reddy et al., 2010).
- Spray chlorpyriphos 20 EC on the trunks and foliage twice, immediately after rains and a fortnight later.
- Spraying of fish oil rosin soap @ 2.5 g / l or carbaryl 50 WP @ 2 g/l is also effective in controlling the pest. (David and Rama Murthy 2016).

Metanastria hyrtaca

This is a minor pest on moringa, generally called gristly citrus caterpillar. It is found all over the Indian sub-continent. This is also a leaf eating caterpillar on moringa.

The eggs are spherical and pale white in colour. Full grown larvae are cylindrical, greyish – brown in colour, stout and hairy. Pupa is brownish red in oblong cocoon usually spun around a branch. Adult exhibits sexual dimorphism. Male moths have pectinate antennae and chocolate brown patch in the middle of forewings. Female moths are bigger in size with longer and broader wings having wavy transparent bands.

Incubation, larval and pupal periods last for 9-12, 45-100 and 9-10 days respectively. The pest has only one life cycle of 75-110 days in a year.

Caterpillars are nocturnal in feeding habit. During day time they remain crowded on shady portion of the tree trunk. They feed gregariously on foliage.

Taragona siva Lef (=Streblote siva)

This is a minor pest on moringa and is widely distributed all over Indian sub-continent. This insect is known as a pest of *Acacia arabica* (Lam) Willd, mahogamy (*Swietensia* spp.), *Rosa* spp. Indian ber (*Ziziphus jujube* Mill), *Polyalthia longifolia* (Sonn) and *Tamarix gallica* L (Fletcher, 1919). A sporadic pest on *Prosopis juliflora* (SW) DC and moringa. *Rosa* spp. is its preferred host.

Damage

Larvae feed on leaves in the early stages of development and remain in groups on the tender shoots. The glistening surface of the leaves and tender shoots indicates the presence of the pest on the tree. Caterpillars feed by scraping of the bark and gnawing foliage, resulting in defoliation. The wooly caterpillar loves the gum of the moringa, the host tree (Beaulah, et al., 2010; Reddy et al., 2010). The grown up larvae are usually found on tree trunks and at times at the base of the main trunk.

The larvae are more or less flat, greyish-brown in colour with long lateral tufts of achreous hairs. Full grown caterpillars are pale chreous brown in colour with small black spots. Full grown larvae form yellowish elongated cocoons on the tree itself and pupate. Adult is stout, chocolate brown moth. Moth has a greyish white head and thorax and whitish abdomen. Forewings are beautifully coloured with reddish brown spots ringed with white border. Hind wings are white with slight fuscous on outer margin. The pest appears from July to November.

Control and Management

- Collect and destroy caterpillars from the plant
- Use light traps to attract and kill adults
- Spray carbaryl (1 kg) or malathion 50 EC or endosulfan 35 EC (1l) in 500-750 litres of water per hectare.

Gram Caterpillar

Heliothis armigera **A**

(Lepidoptera: Noctuidae)

The gram caterpillar feeds on cotton, tomato, moringa etc., and is distributed all over the Indian subcontinent. It's incidence was observed during December – January (Pugalendhi et al., 2010).

Fully grown caterpillar is slender with scattered short hairs. Variable in colour, usually pale apple green with whitish longitudinal lines with a darker shade along the side narrowly edged below. The pupation occurs in brown pupal cocoon in the soil. The adult is medium sized pale brown moth with a distinct double waved anti medial lines on the forewings and white hind wings having a broad blackish outer border (Pugalendhi et al., 2010).

Ash Weevils

(Coleoptera: Curculionidae)

Weevils are categorized as leaf feeders (Ponnuswami, 2012), while Kotikal and Mahesh Math (2016) clubbed weevils with beetles. However, they also feed on roots and pods. Subramanian (1965), observed five species of weevils causing damage to the leaves of drumstick plants at Coimbatore and described them with host range and damage (Table 22.2). Honnalingappa (2001) recorded two weevils at Bangalore (Table 22.2). Mahesh Math and Kotikal (2014) reported three species of weevils feeding on drumstick trees from Bagalkot, Karnataka (Table 22.2). Of late 5 species of weevils were mentioned in their review by Joshi et al., (2016) (Table 22.2).

Table 22.2 List of weevils reported on drumstick in India

S.No.	Weevil species	Place	Reference
1.	*Mylocerus viridanus* Fab	Coimbatore	Subramaniam, 1965
2.	*M.teniclavis* var. *infertior*	-do-	-do-
3.	*M. pusterlatus* Fot	-do-	-do-
4.	*M.discolour* var. *variegatus*. Boh	-do-	-do-
5.	*Ptochus ovulum* FST	-do-	-do-
6.	*M. viridanus* Fab	Bangalore	Honnalingappa, 2001
7.	*Airlanthis*	Bangalore	-do-
8.	*M. viridanus* Fab	Bagalkot	Mahesh Math and Kotikal, 2014
9.	*M. discolor* Rohenman	Bagalkot	Mahesh Math and Kotikal, 2014
10.	Unidentified sp. of *Mylocerus*	Bagalkot	Mahesh Math and Kotikal, 2014
11.	*M.discolor* var. *variegatum* Boh	-	Joshi et al., 2016
12.	*M.undecim pustulatus maculosus* Desh	-	Joshi et al., 2016
13.	*M.teniclavis* var. *inferior* Marshall	-	Joshi et al., 2016
14.	*M. viridanus* Fab	-	Joshi et al., 2016
15.	*Ptochus ovulum* FST	-	Joshi et al., 2016

Adult weevils cause appreciable damage to the leaves of drumstick, while the grubs of weevils feed on the roots of some cultivated crops including drumstick and grasses etc. (Plate 22.7). The weevils feed on young plants and freshly planted limb cuttings (stumps). The weevils notch the edges of the leaves. Adults congregate on tender leaves, mostly on the under surface and nibble the leaves, starting from the margins and working towards the mid-rib, finally consuming the entire leaf lamina (Mahesh Math and Kotikal, 2014). The adult weevils cause notching of leaves, while grubs feed on roots and cause wilting of plant (Usha Rani, et al., 2010). In case of severe attack the weevil scraps the surface of the pods and causes exudation of gum from the pods, which affects the quality of pods (Subramanian, 1965).

Fig. 22.7 Weevil on moringa leaf.

Biology

Eggs are laid in the soil, where pupation also takes place. Adults come out of soil and feed on leaves. Generally weevils appear throughout the year. The peak incidence varies with different species of weevils. For example *Mylocerus* II – *Pustulus* occurs in good numbers during the months of November and December. On the other hand, *M.discolor* var. *variegatum* Boh occurs in good numbers on moringa from January to March. Whereas the adults of undetermined species of *Mylocerus* were observed in the months of July to December. No incidence occurred during second fortnight of December to first fortnight of February (Subramanian, 1965).

Control and Management

- Collect and destroy the adult weevils
- Apply carbofuran 3 G at 15 kg / ha at 15 days after planting (Ponnuswami, 2012)

Moringa Bud Worm

Noorda moringae Walker

(Lepidoptera: Pyralidae)

Moringa bud worm (*Noorda moringae*) was first described by Cherian and Basheer (1938) from Tamil Nadu, India (Plate 22.8). It is a destructive pest recorded on moringa. It is categorized as the major pest of drumstick, particularly in South India. Moringa is the only known host of the bud worm. Among the different accessions of moringa maintained at Horticulture Research Station, Periakulam, Tamil Nadu, cv. PKM-2 was found to be the preferred host harbouring more number of generations of bud worm (Selvi and Muthukrishnan, 2009).

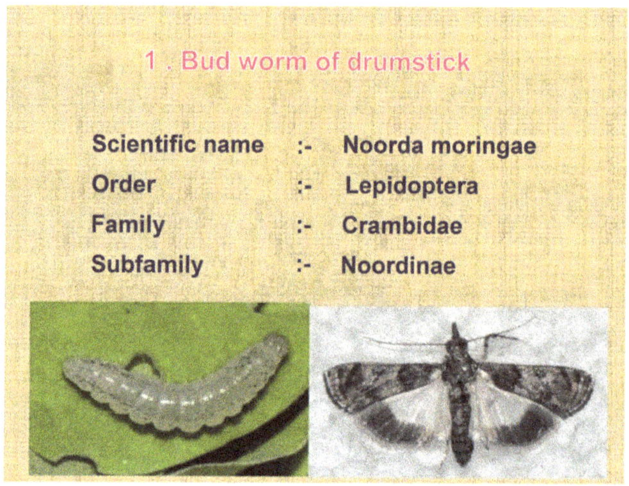

Fig. 22.8 Moringa bud worm.

Damage

Unopened buds are the sites of egg laying, which with a hole are indications of the pest attack, Hatched larvae bore into flower buds, and feed in-side on flower parts. The larvae first feed on anthers and then on the other internal parts leaving outer most sepals and petals intact (Saha et al., 2014). This feeding causes drying and shedding of buds upto 75-78% (Usha Rani et al., 2010). Generally infested buds harbour only one caterpillar. Damaged buds seldom blossom, but dryup and fall down pre-maturely resulting in unfruitfulness of the tree.

For the first time, bud worm was noticed feeding on pods also. Infested pods had irregular holes. Caterpillars bore into the pods and feed on the pulp and seeds. The pest scrapped the pods, causing gummy exudation from the pod, which is considered as the symptom of damage by the pest (Mahesh Math and Kotikal, 2014).

Periods of Incidence

The incidence of bud worm was recorded throughout the year except during November and December in Bangalore (Honnalingappa, 2001), and during December in Bagalkot (Mahesh Math and Kotikal, 2015). The peak infestation of the pest was during April-June under the conditions of Bangalore (Honnalingappa, 2001). At Bagalkot, maximum incidence was recorded in the second fortnight of February and minimum in January (Mahesh Math and Kotikal, 2014, 2015). The difference in the season of peak incidence was attributable to the variation in the weather variables between the locations and also the variety used in the studies. Activity of the pest was more during summer months in Tamil Nadu (Selvi and Muthukrishnan, 2009). At Dapoli (Maharashtra) the incidence of bud worm was noted from second fortnight of October till the end of January and the peak incidence reached in December with gradual increase from second fortnight of October to first fortnight of December, which decreased to minimum during the second fortnight of January (Munj et al., 2001)

Bionomics and Life Cycle (22.9)

Creamy white oval shaped eggs are laid singly or in groups on upopened young flower buds. Hatched larvae are dirty brown with a prominent mid-dorsal stripe and black head and prothoracic shield. Full-fed caterpillars come out of flower buds and pupate in the minute brownish cocoons. Pupae are obligate type and dark brown in colour. Pupation takes place in minute earthen / silken cocoons in the soil or on the ground itself below the dried leaves and debris of the tree basins. Adults are small moths with dull / dark brown forewings and white hind wings having dark brown border (Selvi and Muthukrishnan, 2009).

The life cycle of bud worm is as follows

Egg period	:	3-4 days
Hatching ability	:	36-77%

Larval period	:	8-16 days
Pupal period	:	6-10 days
Adult emergence	:	6-9 days
Total life cycle	:	10-28 days

Control and Management

Cultural Measures

* Ploughing around and below the trees to expose and kill pupae by birds

* Collection and destruction of damaged and fallen flower buds along with caterpillars

* Use of light traps @1-2 per ha to attract and kill the adults in poisonous baits.

Chemical Control

* Since the pest pupates in the soil, apply phorate granules @ 5 g per plant and incorporate into the soil.

* Soon after initiation of flowering, spray sevin (carbaryl) 50 WP (5g/l) or malathion (2 ml/l) or endosulfan (2 ml/l) during October to February (Selvi and Muthukrishnan, 2009).

Biological Control

Natural enemies of the bud worm include the following larval parasites viz., *Pristomeras* spp. (Ichneumonidae), *Bracon brevicornis, Chelonus* spp. (Braconidae), *Elasmus hyblgeae, Perilampus* spp. and *Stytasis* spp. (Chalcidiodae).

Bud Midge

Stictodiplosis moringae **Mani**

(Diptera: Cecidomyiidae)

Bud midge is a minor pest on moringa. Usha Rani et al., (2010) found larvae of bud midge feeding on the internal content of flower buds especially on ovaries, causing shedding of buds in large numbers resulting in barrenness of the trees.

The pest appears during August to January but heavy from October to January (Cherian and Basheer, 1938). Eggs are laid and thrusted in clusters on anthers within the flower buds. Full fed maggots come out of the buds and pupate in the soil beneath the trees. Egg, maggot and pupal periods lost for 1-2, 6-9 and 5-8 days respectively. Adult is free-living brown coloured small insect. The pest has only one generation in a year and completes its life cycle in 12-19 days.

Management and Control

Management and control measures are similar to bud worm (*Noorda moringae* Walk)

1. Plough the soil around and below trees to expose and kill pupae
2. Collect and destroy damaged and dropped flower buds along with maggots.
3. Set up light traps @ 1 per ha to attract and kill the adults.
4. Spray insecticides like carbaryl 50 WP @ 1 g/l or malathion 50 EC @ 2 ml/l of water.

Pod Borer / Fruit Fly

Gitona distigma **Meigen**

(Diptera: Drosophilidae)

Moringa pod borer (fruit fly, pod fly) is an important pest of drumstick and assumed a major pest status in South India (Plate 22.9), as it attacks the fruit (economic part) and causes heavy losses ranging from 72 to 85 per cent during fruiting (Sivagami and David, 1968; Kader and Shanmagavelu, 1982; Anjaneyamurthy and Regupathy, 1989; Reghupathy et al., 1991; Ragumoorthy and SubbaRao, 1997). Pod fly is a Paleaerctic species reported for the first time in 1968 from India (Sivagami and David, 1968). Now it is a serious pest on annual moringa resulting in 80-100 per cent yield loss in neglected orchards under poor management conditions (Ragumoorthy and Armugam, 1992).

The problem of fruit fly increased from the introduction and start of cultivation of annual drumstick. Hence it was presumed that with the advent of annual drumstick, the fruit fly, hither to a minor pest on moringa has assumed major status (Kader and Shanmugavelu, 1982; Anjaneyamurthy and Regupathy, 1992).

Earlier, pod fly was reported as a pest of drumstick and its varieties in different agro-climatic zones of Tamil Nadu (David and Kumaraswamy, 1982; Anjaneyamurthy, 1985).

Period of Incidence

Drumstick is more or less a continuous bearer of fruits. Hence, the incidence of pod fly occurs almost throughout the year, whenever there is fruitset and fruiting on the tree. Incidence, its season and peaks however, are location and associated weather dependent and hence variable.

Abdul Kareem et al., (1974), recorded the incidence of the pest during June to July at Coimbatore. Whereas, Anjaneyamurthy and Regupathy (1988), reported maximum incidence (49.4%) during November to December at Coimbatore. In a later study they observed the incidence almost throughout the year and recorded peak incidence in the months of August – September: Incidence decreased to 13-20% in November-December, subsequently it increased (23.4%) in January – February, which was attributed to lower level of fruit setting during these months, which boosted the percentage of incidence even though only few fruits were affected. Contrary to this was the fact that in spite of heavy fruit set in April-June, the damage was less (0.8 to 2.2%). In other words, less the fruiting more is the damage. Further in summer the incidence as well as damage, both are less probably due to higher temperatures. Another decrease in the incidence was recorded in the months of March to June (Anjaneyamurthy and Reghupathy, 1992).

Mahesh Math and Kotikal (2014, 2015) investigated the seasonal incidence of fruit fly on annual moringa (cv. Bhagya) at Bagalkot (Karnataka). The maximum incidence (35% pod damage) by the fruit fly was noted during second fortnight of December and minimum (21.7% pod damage) in early stage of pod formation during second fortnight of October. There was no damage from February to May.

The activity of the pest was more from April to October, which declined there after in Bihar (Saha et al., 2014).

Ecology

Weather conditions / variables (temperature, relative humidity etc) exert marked influence on the incidence of insect pests including pod fly on moringa at different locations of its growing. Anjaneyamurthy and Reghupathy (1992) at Coimbatore observed that the incidence of fruit fly was negatively correlated with maximum temperature and hours of sunlight and positive correlation with relative humidity and sunshine hours of previous months. Thus the variation in the incidence of fruit fly may be attributed to the variation in the weather parameters not only in the current period, but also in the previous months of incidence.

At Bagalkot (Karnataka) the incidence of fruit fly showed highly significant and negative correlation with maximum temperature (r=0.8713) and non-significant correlation with minimum temperature (r=0.0.408) and positive correlation with evening relative humidity (r=0.04222) and total rainfall (r=+0.017) (Mahesh Math and Kotikal, 2015).

Damage and Symptoms

Flowering in moringa starts from May in Tamil Nadu and during this period the incidence of fruit fly also starts and persists till harvesting stage. Adult female lays eggs in the groves / cavities of fruits at the distal end. The hatched larvae (maggots) bore holes for access into the fruit. The entered maggots feed on the pulp and seed of the fruit. The boring and ingress of larvae cause gum oozing from the affected fruits. As a result of pest attack the fruit becomes thin, twisted and dries up from the tip of the fruit upwards, upto the base of the fruit stalk. As a result of fruit fly attack, the fruit growth is arrested (Honnalingappa, 2001). Internal contents of the pods rot. Decaying of the affected fruits attracts other saprophytic flies and microbes and the infested fruits get putrefied and emit a very characteristic odour (smell). The damaged fruits are totally rendered unfit for human consumption leading not only to yield loss (60-65%) but also economic loss to the farmer (Abdul Kareem et al., 1974; Regupathy et al., 1991). The extent of fruit damage ranged from 0.8 to 49 per cent (Anjaneyamurthy and Regupathy, 1992).

Ragumoorthy (1996) reported that _Gitona distigma_ the fruit fly, caused 70 per cent yield loss in Tamil Nadu, because of which it attained a major pest status. The yield loss is still higher in poorly managed moringa plantation (Ragumoorthy and Armugam, 1991, 1992). Economic injury level (EIL) is 15 per cent of affected fruits (Ragumoorthy et al., 1998). A maximum of 20-28 maggots may be found in a single fruit.

The characteristic symptoms of attack and damage can be observed from the initial stages of infestation of the pest by the presence of gum exudation in association with egg laying and by drying of pods at later stages (Beaulah et al., 2010). The drying and splitting of fruits are also indications of pest incidence (Honnalingappa, 2001).

Bionomics and Biology (Plate 22.9)

Eggs are cigar shaped, sculptured and white coloured and are laid singly or in groups of 3-4 in fully opened flowers or in groves of developing fruits. Eggs are glued in groves between the ridges of pods. Maggots are cream coloured. Full grown maggots come out of the fruits and fall to the ground for pupation (Pugalendhi et al., 2010). Adults are small in size, yellowish brown in colour with transparent wings that extend beyond the body with two black spots on the forewings near the coastal margins and red coloured compound eyes. The female adult after emergence from soil and mating,again attacks the fruits for egg laying and thus repeats the cycle (Regupathy et al, 1991).

Fig. 22.9 Adult pod fly.

The biology of the pest is as follows:

Egg Period	:	3-4 days
Maggot period	:	18-25 days
Pupal period	:	5-9 days

Control and Management

Several moringa entomologists worked on the management and control of fruit fly on moringa and suggested both cultural and chemical control measures. Integrated pest management schedules were also tested and suggested. Various management practices have been tried from poisonous baits to application of pesticides with varying success (Ragumoorthy and Armugam, 1992; Mohan et al., 1993; Mahesh Math et al., 2014a).

Cultural Measures

Management / control of pod fly on moringa is possible with regular non-pesticide cultural measures with or without insecticides, provided the control of fruit fly is started when 1 or 2 affected fruits are observed on the trees.

- Collect periodically fruit fly damaged moringa pods and destroy them by dumping in a pit, (3′ deep) and covering the pit with a thick layer of soil to prevent the emergence of adults and carry over of the pest.

- Frequently rake the soil under the trees or plough the infested field to destroy the puparia

- Do not allow water stagnation in the tree basins and / or in the field. Keep dry and clean, since under water stagnation, pest develops well, besides the incidence of root rot.

- Do not allow weeds on the field bunds as well as in the field.

- Arrange pasonous baits with palmyrah sugar or cane sugar or jaggery (1%) plus 5 ml of malathion (1%) in each bait to attract the flies and kill. Methyl eugenol is commonly used in poisonous traps. However, as moringa pod flies are not attracted to methyl eugenol and fish meal, use attractants like citronella oil / eucalyptus oil / vinegar (acetic acid), dextrose or lactic acid to trap flies and to kill (Saha et al., 2014).

- Fruit flies can be effectively managed (checked) by prophylactic sprays of neem based pesticides. These sprays also check further egg laying as a bonus. Powdered neem cake and lebacid check further infestation.

- Soil application of neem seed kernel extract (NSKE) at 2 litres per hectare and lindane 10 D at 200 g per tree during 50% fruit set reduces the incidence of fruit fly (Ragumoorthi, 1996).

Chemical Control

- After the destruction of affected pods, spray monocrotophos (1.5 ml/l) or desis (1 ml/l) or carbendazim (1 g/l) twice at 10-12 days interval (Madhusudan Reddy, 2002) or Spinosad (Beaulah et al., 2010).

- To destroy the pupae in the soil, dig the soil to dry and then in corporate phorate granules (5 g per tree) or neem cake powder (500 g per tree) or BHC 10% dust (Madhusudan Reddy, 2002). At a later stage spray fenthion (0.04%) or dichlorvos (0.4%) or monocrotophos (0.05%).

- Kathiresan et al., (1999) advised spraying of fenthion (1 ml/l) and repetition 15 days after based on the need to control the pest.

- It was suggested to apply lindane at 25 kg/ha after racking the soil below the trees (Beaulah et al., 2010).

- It was recommended to spray dichlorvos 76 SC (500 ml) or malathion 50 EC (750 ml) in 500-750 litres of water per hectare, when pods are 20-30 days old and to apply Azadirachtin (0.03%) during 50% fruit set and 30 days later.

- Application of Emamectin benzoate 5G at 0.25 g/l and Spinosad 45 SC at 0.20 ml/l were superior to control fruit fly on moringa at Bagalkot (Mahesh Math et al., 2014).

IPM schedule

A. Tamil Nadu Agricultural University, Coimbatore developed the following IPM schedule for management of moringa fruit fly (Regupathy et al., 1991).
  - Collect all the infested fruits and destroy
  - Keep the field weed free to prevent further spread and multiplication of fruit fly
  - Stir the soil beneath the free canopy and incorporate BHC 10% dust during fruit maturity.
  - Spray either dichlorvos (0.4%) or neem cake extract (1%) or fenthion (0.04%) twice-once at 50% flowering and next 15 days later.

B. The following IPM was suggested for effective management of fruit fly incidence on moringa (Pugalendhi et al., 2010).
  - Weekly removal of damaged fruits and their destruction
  - Soil application of NSKE @ 2.0 l/tree at 50% fruit set
  - Spraying of fenthion 80 EC @ 0.04% during vegetative and flowering stages
  - Spraying of nimbicidine (0.03%) 150 ppm during 50% fruit set i.e. 35 days after flowering.

C. Sreenivasa Rao and Hari Prasad Rao (1998) have suggested the following spray schedule for the control of drumstick pod fly in Andhra Pradesh.
  - At 25% flowering – spray zolone (2 ml/l)
  - At 50% flowering – Dimethoate (2 ml/l)
  - At 100% flowering – Monocrotophos (1.5 to ml/l)

The benefit of IPM would be: efficient control of pests and substantial increase in yield returns. Muthukrishnan (2009) obtained increased pod yield and higher benefit: cost ratio in IPM module-I having various components including Spinosad and profenophos. Mahesh Math et al., (2014) realized higher marketable pod yield with spraying of Emamectin benzoate and profenophos.

Harvesting Precaution

Since chemical pesticides are toxic to humans, it is advised to harvest edible mature pods of drumstick before spraying for markets and / or after application of pesticides after a waiting period of 10-15 days (depending on the pesticide applied).

Mites

1. *Tetranychus neocaledronicus* (Andre) (Spider mite)
   (Acarina: Tetranychidae)
2. *Aculus moringae* Channabasavanna
   (Acarina: Eriophyidae)
3. *Aculus pterigospermae* Keife
   (Acarina: Eriophyidae
4. *A. culusmenoni* Channabasavanna
   (Acerina: Eriophyidae)

Mites are non-insect sucking pests. They are small arachinids, relatives of ticks and spiders. They are easy to see on plant leaves because of a combination of three characteristics (a) mites can be seen as tiny red or black dots on the underside of the foliage, (b) webbing all over the leaves and terminal twigs, and (c) speckled bleaching of the leaves.

Butani and Verma (1981) reported two species of mites, while Honnalingappa (2001), listed four mites invading moringa in India. *Tetranychus neocaledronicus* (Andre), the spider mite was recorded by Banu and Channabasavanna (1972) and Sangeetha and Ramani (2007). *Aculusmenoni* was reported by Butani and Verma (1981), Mohan Sundaram (1985) and Honnalingappa (2001) mentioned the *Aculus pterigospermae Keifer* on moringa. *Aculus moringae* was reported by

Butani and Verma (1981), Mohan Sundaram (1985) and Honnnalingappa (2001).

Mites enjoy worldwide distribution on several economically important agricultural crops. Mites are the largest economic pests of moringa leaves in Hawaii because of which there is possible rejection of shipments of moringa leaves and pods to North America (Radoivch, 2009).

Tetranychus neocaledonicus (Andre) (Vegetative / Spider mites)

Spider mites are so called because of the cobweb like webbing left behind on the foliage. Moringa is the favourite host of mites. However, susceptibility of different *Moringa* spp differs. Mites are not a trouble with *Moringa peregrina* but *M.concanensis* and *M.rivae* are susceptible, whereas *M.borziano* is not susceptible. However, mites are very fond of *M.oleifera* Lamk (Olson, 2014).

Aculus menoni and *A.moringae* are vigrants on both surfaces of leaves and on leaves and stems respectively, causing no apparent symptoms on the hosts (Sangeetha and Ramani, 2007).

Ecology

Environmental variables especially temperature and relative humidity (RH) have significant bearing on oviposition and fecundity of spider mites. Compared to temperature, RH exhibited little influence on pre- and post-oviposition. Fecundity was highest at 34 ± 1^0C and $50\pm5\%$ RH. Higher RH showed negative impact on egg laying capacity of the mites. At low temperature and higher humidity a decrease in the duration of oviposition period was observed (Beaulah et al., 2010). Mite incidence would be severe when the trees are stressed. Mites in general increase during dry and cool weather, causing damage to the plants, which recover during warm weather (Palada and Chang, 2003).

Damage

Mites feed on leaves by sucking sap from under side of leaves. Spider mites feed by puncturing the leaves / tender plant parts with the help of stylets (Yousuf and Chouhan, 2009). The infested leaves turn whitish first as chlorophyll is lost and later turn yellow (Sangeetha and Ramani, 2007) which are shed subsequently (Plate 22.10 and 22.11).

Fig. 22.10 Red spider mite damage of moringa leaves.

Fig. 22.11 Red spider mite damage of moringa leaves.

Management and Control

There are several smart ways to manage and get rid of problematic mites on drumstick.

- Do nothing – infested leaves soon drop along with mites. Dropped leaves may be collected and destroyed by burning to kill the mites.

- Spray mite infested moringa plants regularly with strong jets of water particularly on the underside of the leaves to knock out adult mites and their webbing off the plant. When done often –once in a week or two, depending on the intensity of incidence, mites could be kept at bay (Joshi et al., 2016).

- Best strategy for large and vigorous moringa trees that have mites or any other pest is, simply cut back the whole tree back to a height of 150 cm or even close to the ground. Next strip the infested leaves and stems and destroy by burning.

- Use oils like neem oil diluted with water, which can effectively kill spider mites. Unlike synthetic pesticides, oils offer several benefits, they can effectively and safely be used on fruits, vegetables etc., not harmful to beneficial insects. The oils can interrupt the reproductive cycle of mites and kill eggs. Neem oil is a preferred natural product that can be successfully used against spider mites and upto the day of harvest unlike synthetic chemical pesticides. Neem oil kills all stages of spider mites including eggs and young ones (Anon, 2014b).

- When it is necessary to employ synthetic pesticides choose the correct pesticide that can kill not only live mites but their eggs as well. For optimum results spray the plants at correct time. Pesticides like imidacloprid (systemic) for thrips control can make mite infestation worse by killing the beneficial insects like mite-eating mantis on the plants. So use acaricides instead of insecticides (Anon, 2014b).

Vertebrate Pests

Besides non-vertebrate pests like insects and mites, vertebrate pests like cattle, sheep, goats and pigs feed on moringa seedlings, leaves and pods. In North India owl birds damage the flowers of moringa by sucking the sap from flowers (Verma and Chaarasia, 1993). In Africa, antelops, rats etc eat seedlings, leaves and pods. Vertebrates love drumstick very much, particularly seedlings in nurseries and leaves on grown up plants.

In view of the above, protection of moringa seedlings, leaves and pods is a great concern at times. The seedlings may be protected by installing a fence, until they grow up as trees and more mature. In Nigeria fences made out of the branches pruned from moringa trees are built around the nurseries as a protection (Mridha, 2015). In Africa moringas are protected by installing a fence or planting a hedge around the plants. A live fence with *Jatropha curcas* is planted to protect moringa seedlings from cattle etc. In case of mature trees lower branches can be pruned at a height so that goats will not be able to reach the leaves and pods for grazing (Verma and Chaarasia, 1993).

Integrated Pest Management (IPM)

Moringa (*Moringa oleifera* Lam.) is invaded by several pests both insects and non-insects (mites). However, it is difficult to adopt separate and individual plant protection measures for each and every pest and is not feasible and practicable. Further different pests appear at different times with varying intensities. Hence, integrated pest management (IPM) has been suggested by which most of the moringa pests can be controlled. IPM includes all types of management practices, cultural, chemical and biological measures individually as well as in combination. The following IPM measures for the management and control of moringa pests have been suggested.

- Use nitrogen fertilization moderately which prevents most of the pests and diseases.

- Apply fertilizers in balanced quantities. At last ploughing apply FYM (10 t), single super phosphate (100 kg) and top dress thrice (at second, fourth and sixth month after planting). At second month after planting 250 g per tree castor or neem cake + urea (60 g) + muriate of potash (30 g). In fourth and sixth months after planting apply urea (60 g) + muriate of potash (30 g) per tree.

- Do not allow water stagnation in the field particularly in the tree basins since in water stagnation conditions fruit fly multiplies in excess numbers and roots rot.

- Do not allow weeds in the field or in the tree basins or even on field bunds, since some of the weeds serve as alternate hosts of certain insect pests.

- Several of the insect pests pupate / hibernate in the soil. Hence, now and then stir the basins or plough the field to expose them to predation by birds and to the Sun for destruction.

- During day time leaf eating caterpillars congregate at the base of the tree trunk seeking shade, in large numbers particularly in case of perennial types of drumstick. Arrange slow burning fires near them when their congregation was observed or expose them to burning torches to destroy the caterpillars. This is a cheap and feasible method.

- In the borer holes, after cleaning them inject a mixture of dichlorvos and water (1:1) or petrol or carbon disulphide or chloroform or celphos tablets and plug the holes with wet mud.

- Incorporate in each tree basin, 50-70 g of carbofuran (3%) granules or 20 g phorate (10%) granules (without touching the trunk) to kill pupae in the soil atleast twice in a year at appropriate time of pupation.

- Arrange poisonous baits in the field at 1 or 2 per hectare. Baits are to be prepared with a mixture of palmyrah sap / cane sugar / jaggery in water at 1:1 ratio plus malathion (5%) 5 ml in earthen saucers and arrange them here and there in the field to attract adults of fruit borers and flower borers etc., which get attracted, feed on the poisonous bait and die.

- Apply fenthion 80 EC (0.04%) during vegetative and flowering stages.

- At the time of flowering incorporate Acephate in the soil. During flower formation and bud development stages spray Zolan (2 ml/l), at 50% flowering spray dimethoate (2 ml/l) + Neem oil (5 ml/l) and finally at 100% flowering spray monocrotophos (2 ml) + neem oil (5 ml/l).

- For managing leaf eating caterpillars, before flowering apply endosulfan (2ml/l) or quinolphos (2 ml) or monocrotophos (2ml).

- After 50% fruit set spray nuvon (1 ml) or dimethoate (2 ml) or endosulfan (2 ml) + neem oil (5 ml). After 30-35 days later apply neem seed kernel extract to soil @ 2 litres per tree along with the weekly removal of damaged fruits.

- Observe the time of egg laying of bud borer and spray monocrotophos (2 ml) to control the pest.

Precautions

- Although there are no / few reports on variability in the pest tolerance with in moringa germplasm, local varieties are most likely to be best adapted to local conditions and local pests and diseases so it is advisable to include them to the extent possible in newer plantation (Beaulah et al., 2010).

- Use low impact, neem oil, horticultural soaps and sulphur to control sucking pests like aphids, mites etc. (Beaulah et al., 2010).

- Chemical control of insect pests may be used when severe infestations occur. Choose a pesticide that targets the specific pest(s) causing the damage and avoid pesticides that kill or inhibit the development of beneficial organisms. Choose pesticides that lost only a few days (Palada and Chang, 2003).

- As a caution donot use the same chemical twice / every time, change the chemical every time for effective control of the pest.

- Harvest all edible mature pods before spraying pesticides.

Pest Surveillance (For export)

Moringa pods are exported from India to different countries. The importing countries need pest free pods. Hence it is necessary to adopt pest surveillance for obtaining pest free pods of moringa. The procedure is as follows:

- Weekly monitoring should be done through pest scouting with the help of monitoring devices like pheromones and coloured sticky traps.

- For field scouting 300 fruits from 100 plants per acre should be observed. Minimum 15 spots at reasonable distance with each other following a cross diagonal pattern moving in zig-zag manner for counting all types of insect pests should be observed.

- Pest monitoring for fruit flies using traps should be done regularly from fruit set stage onwards.

- If 95% plants are found free from insect pests, then the field will be considered fit for export.

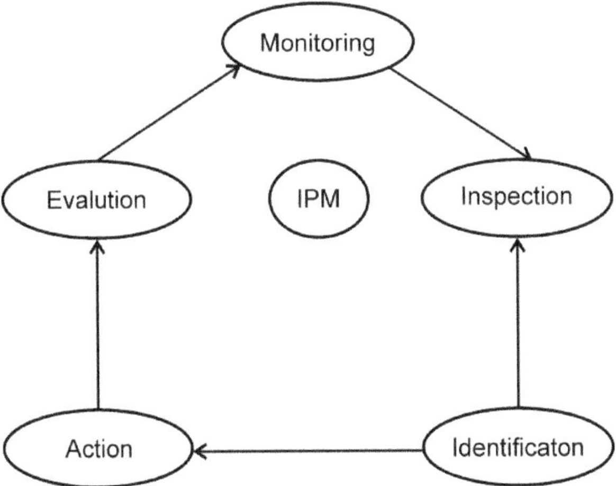

Fig 22.12 Important activities for pest free drumstick production to export

CHAPTER 23

Off Season Production

Production of crops should be according to the market demand / follow the demand in the market. Moringa enjoys market demand for pods throughout the year in the country. Though drumstick provides two crops a year, its demand is very high during winter particularly in Tamil Nadu as the winter period coincides with festivals and marriages, wherein a meal is considered incomplete without moringa Unfortunately production of pods during this period (winter) is less.

During the north-east monsoon period (October-December), prevailing low temperature both in soil and in the atmosphere and high soil moisture content affect the flower production in moringa (Bharati and Pugalendhi, 2014). Further, heavy rainfall and low temperature during winter lead to the drop of flowers resulting in poor fruitset and production. This period, therefore is considered to be the lean or off-season for moringa.

During this season for want of adequate production and high demand, the prices of the pods will be higher and even rise up to Rs.60-100 / kg of pods which is higher than the price i.e. Rs.5/kg of the main season (March to August). Accordingly the profits obtained during the glut period (March to August) will be higher than during winter due to higher production. Even if 50% production is achieved during the lean period farmers will be getting as high as 4-6 lakhs of rupees per hectare as income which is un-imaginable from other vegetable crops.

In view of the above, attempts were made with following approaches to induce off-season flowering and pod set during the lean period (November-February) especially in Tamil Nadu.

(i) Staggering seed sowing, followed by spraying chemicals / plant growth regulators.

(ii) Pruning the trees followed by spraying chemicals / Plant growth regulators.

(iii) Mulching followed by canopy management by pruning and chemical manipulation by growth retardants like uniconazole

I. Staggering of Seed Sowing dates

Generally, annual moringa seedlings take 5-6 months-time to come to flowering from the date of sowing. Hence, it was conceived that by staggering the dates of sowing of seeds, it is possible to achieve off-season production in moringa. An attempt made at Periakulam, Tamil Nadu demonstrated the possibility of obtaining off-season production in moringa with staggering of sowing dates followed by chemical treatments (Pugalendhi and Swaminathan, 2010).

Seeds were sown in May, seedlings were pinched twice, once when they are two feet high and next 25 days later, which resulted in the development of umbrella shaped tree with many branches. Then the trees were sprayed with different chemicals during August-September to induce flowering. The combination of May sowing plus foliar spray of nitrogen (urea 0.5%) resulted in off-season production during November-February with early flowering, increased number of pods per tree, and highest yield of 8.58 t /ha (Pugalendhi and Swaminathan (2010).

II. Pruning Trees

Pruning one-year old trees, after heading back in July along with the application of potassium nitrate (0.5%) and application of 10 kg FYM along with urea super phosphate, muriate of potash @ 100, 100 and 50g per tree influenced positively in giving off-season production. These trees recorded higher number of panicles per tree, number of flowers per panicle and total number of pods per tree, pod weight and yield per tree.

III. Mulching the Trees

Higher soil moisture and low soil temperature hinder flowering during winter in moringa. Hence it is conceived to reduce soil moisture content and rise the soil temperature with mulching. Studies of Bharati and Pugalendhi (2014) indicated the possibility of off-season production in moringa with black polythene mulch (50 micron thickness).

Mulching was done during August to March to prevent the entry of rainwater into the soil and increase soil temperature around the plant. Such mulched trees were pruned after one year of planting in the month of July on which growth retardant uniconazole (50 ppm) was sprayed one month after pruning. This resulted in off-season production with good pod characteristics. The possibility of off-season production by above treatments was attributed to increasing soil temperature from 2.2 to 3.4^0C during north east monsoon period under plastic mulch and favourable effects of pruning as well as growth retardant viz., uniconazole (Bharati and Pugalendhi, 2014). Therefore, pruning in July and August combined with mulching and application of unicanozole (50 ppm) would be ideal to achieve off-season production of annual moringa cv. PKM-1.

Singh (2011) suggested a new approach of withholding irrigation followed by pruning to induce off-season production in moringa. Probably, taking a cue from the above suggestion, one Mr. Nader (2018) software engineer turned farmer attempted and achieved off-season production of ODC_3 variety of moringa in Tamil Nadu with the following measures:

(i) He has sown the seeds between 30th April and 15th May

(ii) Pruned the trees when they reached a height of 2 feet and subsequently 25 days later.

(iii) Between 70th and 85th day of planting, gave two sprays of potassium nitrate (0.5%) and nitrobenzene (0.5%)

(iv) During first week of September the trees were manured with 5-10 kg of poultry manure per plant to generate heat in the soil for the induction of flowering.

(v) During the second week of September, trees were given water stress by stopping irrigation or giving less water.

(vi) This resulted in the off-season production, 90-100 days after planting during September-February

(vii) However, it was cautioned that flowers should not be allowed before last week of September. But reason for such caution has not been stated.

CHAPTER 24

Organic Cultivation of Moringa

Organic farming is now increasingly adopted in both agricultural and horticultural (fruits, vegetables, flowers, plantation crops etc) crops. Among the vegetable crops "Moringa" fits well as a candidate crop for organic cultivation. It is fast growing, drought tolerant, hardy, and resistant to most of its pests and diseases, perennial and grows well in all types of soils and capable of adapting to varied eco-systems. In fact moringa has been cultivated purely through organic farming in India during the past several decades and even now in back yards etc. According to Prabhakar and Hebbar (2007), organic production of moringa is feasible and suitable economically as well as socially in the present context of reducing depletion and reducing pollution of natural resources and cost of farm production and encouraging foreign exports. Further, it is clear from their studies that organic cultivation of drumstick using different sources of organic manures alone or in conjunction with others like biofertilizers preserves the natural resources, environment free of chemical pollution and this technology is likely to make a long head way in future in the country. Singh (2011) opined that since moringa is used for food, neutraceuticals etc. it would be essential to use minimum amount of chemical fertilizers and pesticides which is possible by adopting organic farming. In moringa cultivation, organic farming could be adopted through growing green manure crops as intercrops for incorporation in the soil, vermicomposting for meeting nutrient needs of the crop, and IPM produce residue free fruits and leaves, besides the use of organic manures only. Recycling of farm waste, use of micro-organisms (biofertilizers) and vermi-wash would be desirable in organic cultivation of moringa.

Organic manures useful for organic farming of moringa are: farm yard manure, biofertilizer, vermin compost, vermi wash, oil cake, panchakavya and organic pesticides and animal manure. (See Manuring and Fertilization). Since animal manures are suggested for organic cultivation farmers must have animals (Singh, 2011).

Package of Practices for Organic Farming of Moringa

Studies conducted using organic manures, biofertilizers and inorganic fertilizers clearly brought out into light the advantages and beneficial influence of organic manures and biofertilizers in improving the growth, enhancing yields of pods and leaves and betterment of their chemical quality in drumstick over inorganic fertilizers (see manuring and fertilization). But no definite information on package of practices (POP) of organic moringa is available in the literature. However, Save Indian Farmers (SIF) a United States of America based organization working in India suggested the following POP to moringa growers of Maharashtra and Andhra Pradesh in India.

Climate

Moringa requires hot and humid climate for growth and dry climate for flowering and fruiting. The temperature of 25^0-30^0C is suitable for flowering. Area for moringa should be frost free.

Soil

The requirement of soil for organic moringa is principally same as that of general farming. Well drained loam to clay loams are ideal; soil pH should be 6.2 to 7.0. It cannot withstand poor drainage and prolonged water logging in the field.

Land Preparation

The selected land should be prepared well by giving atleast two ploughings. At final ploughing FYM @ 20 t per ha should be incorporated into the soil. Green manure crops may be raised and incorporated into the soil when they start flowering (30-45 DAS) to enrich the soil with the organic matter (humus). After leveling the land, pits of appropriate size may be dug up and filled with top soil mixed FYM and compost.

Varieties

Generally annual moringa cvs. PKM-1 and PKM-2 are preferred for organic cultivation. Other varieties viz., Rohit-1, Coimbatore-1 and Dhanraj can also be selected.

Propagation and Planting

Moringa plantation can be raised either by direct seeding or transplanting the seedlings raised in nurseries (beds / containers). Best time for sowing is either, June-July or January-March. It is advisable to transplant one month-old seedlings at a spacing of 3 x 3 m.

Water Management and Method of Irrigation

Being a drought hardy (tolerant) crop, moringa needs less water, hence it is capable of giving yield in drought conditions also, but the yields are less. Hence, for commercial organic farming it needs irrigation. With proper irrigation and fertigation, it will give more yield. Good yields can be obtained with drip system of irrigation.

Manures and Fertigation

FYM @ 20 t/ha is applied and incorporated in the soil during final ploughing. Dug up pit is filled with 8-10 kg FYM/compost.

Pinching and Pruning

When the seedling / plant reaches a height of 75-100 cm, the terminal / apex shoot of the plant may be removed (pinched / tipped) to allow the emergene of lateral shoots for better productivity and production of yield of pods / leaves. First pruning may be done after 2 months of planting when the plant reaches a height of one metre after pinching.

Weeding

Field should be kept weed free by regular weeding operations. Intercropping with other crops like vegetable can also control weeds.

Pest and Disease Control

Prophylatics such as diluted cow urine and vermiwash are often used for pest control in moringa organic farms. Where prophylatics fail and pest populations reaches peak population and economic loss is inevitable the following non-chemical methods of pest control are advised.

(i)   Hand picking of the pest (where the pest is a large caterpillar / grubs for example)

(ii)   Use of pheromone traps (for controlling fruitfly),

(iii)   Use of light traps (for moths etc)

(iv)   Use sticky traps (for mites)

(v)   Use of predators (natural enemies)

(vi)   Growing trap crops (alternate / collateral hosts)

(vii)   Use of microbial pesticides and biological agents like *Trichogramma, Trichoderma.*

Leaf eating and hairy caterpillars can be controlled with pheromone traps, jassids and mites can be managed with sticky traps installed in the field, bark eating caterpillar can be controlled by inserting iron wire (spoke) into the holes or inserting cotton ball soaked in petrol and closing the hole with mud. Fruit fly (*Gitona distigma*) can be effectively managed by adopting IPM measures which include; i) application of nimbicidine(0.03%) at 150 ppm during 50% fruitset and 35 days later; ii) soil application of neem seed kernel extract (NSKE) @ 2 litres per tree at 50% fruitset and iii) weekly removal of affected fruits and their destruction (see insect pests)

Rotting of roots and stems in water logged conditions can be avoided / managed by draining excess water from the field.

Harvesting and Yield

Fruits (pods) may be harvested at correct edible maturity stage to obtain best quality produce with optimum level of components for human consumption. Yield of pods will be 50-55 tonnes per ha and per tree yields may be 220-250 pods per year.

CHAPTER 25

Future

Interest in Moringa in recent times has been shown towards its nutritional and medicinal properties, hence much research has also gone into this aspect, consequently the demand for its products has been ascending, which should be exploited; by creating general awareness among the people about its benefits, especially with the rural sections. The plant has enormous medicinal value blended with high nutritional supplement which has full power to boost up health of citizens on one hand and on the other it can solve food problems and mal-nutrition in infants (Benerjee and Batra, 2012). Therefore, moringa should be promoted for greater consumption for human use to improve nutrition and strengthen immune systems.

Because of its tolerance to poor soil conditions and hardy nature against-droughts, mechanical injuries, moringa can be planted in degraded soils for soil conservation. It is also recommended for reclamation of mined areas. It is therefore, highly recommended for social forestry programmes. It has tremendous potential in agro-forestry systems as it is a deciduous tree and adds humus by its litter fall and also cast minimum shade on under story crops (Tewari et al., 2003). In view of its multiple uses, the moringa needs to be widely cultivated in most areas, where agro-climatic conditions favour its optimum growth and production. The planting of trees by small holding farmers should be encouraged because it will improve their health and wealth (see success stories).

Moringa is somewhat under-utilized. Therefore, more research work can be done on humans so that a drug with mutli-ferous effects will be available in the future.

Moringa is mainly grown for pods in India. Its leaves are not much used. Adding moringa leaf powder in cooking should be encouraged particularly in developing countries to combat under nourishment and malnutrition.

For increasing area and production in moringa, high yielding varieties and better package of practices are needed for successful domestication and large scale cultivation. For successful production and large scale commercialization of moringa and its related products, conventional, modern breeding and crop improvement programmes should be established for the purpose of improving the yield and quality of its products.

By following the above, India could easily fight against the problem of malnutrition, poverty, disease, unemployment and export of its products to earn foreign exchange by utilizing its full benefits.

Success Stories in Drumstick Cultivation

Following the development and release of improved varieties of drumstick from Tamil Nadu Agricultural University, Periakulam, University of Horticultural Sciences, Bagalkot, Karnataka and by some private farmers, farmers started showing interest in the cultivation of drumstick particularly in drought prone regions of different states of the country and achieved loudable success. Its high demand for pods, precocity in bearing, drought tolerance, less water requirement, free adaptability to different agro-ecological conditions, less investment and quick and high returns have attracted the farmers towards drumstick cultivation. Further, its easy cultivation with less efforts and time also played role in attracting many employees to enter into moringa farming leaving their jobs. Observing the successful drumstick cultivation of other farmers and unsuccessful farming of traditional agricultural and horticultural crops due to drought, decline in water resources, pest and diseases attack, financial constraints, un-remunerative prices and returns etc., some farmers have switched over to drumstick farming. Due to certain undesirable effects and higher cost of chemical farming and long duration of perennial horticultural crops some farmers adopted organic farming of drumstick. Some NGOs and Government also encouraged the cultivation of drumstick by providing information on the cultivation and supplying inputs to the farmers. Some of the success stories of farmers from different states of the country are narrated below:

Andhra Pradesh

Andhra Pradesh is one of the important states in the country and ranks first in area and production of drumstick. "Save Indian Farmers" (SIF) a NGO organization of USA, working in Andhra Pradesh, Telangana and Maharashtra is instrumental in educating and encouraging the farmers to cultivate different crops including, moringa, particularly in drought prone regions of these states. It supplies inputs also for the cultivation of crops. SIF introduced drought tolerant and high yielding varieties of drumstick to few farmers in Prakasam district of Andhra Pradesh with the coordination of one progressive farmer – **Nagireddy Prabhakar Reddy**. As the drumstick yields around the year, the crop has been embraced by the farmers. The cultivation of drumstick is spreading in the state as droughts become a regular feature. Moringa is among the few horticultural crops which begins fruiting within six months of planting and continues to do so for a period of 10-15 years. Farmers of SIF have made more than Rs.2.0 lakhs in the first six months of planting per hectare (Anon, 2019).

Irigala Brahma Reddy, a farmer from Venkatapuram village of Prakasam district with his 3 acres farm, has become a shining example of dry land farming by growing drumstick successfully in a sparse vegetative landscape having 40^0C temperature and squat houses. SIF supplied planting material and drip irrigation equipment and educated him about drumstick cultivation. He has earned a profit of Rs.1.0 lakh and is expecting another Rs.1.0 lakh this year, even in the presence of decline of water in the bore wells due to drought in his area.

Another innovative farmer, **Chakkara Bhupal Reddy** in Prakasam district with the advice of SIF has taken up drumstick cultivation in his 2 acre land with irrigation and inter crops like chillies and beans. As of January 2019, he harvested 2000 kg of drumsticks from his 2 acre plantation and got an income of Rs.1,30,000/- on initial investment of Rs.35,000/- which enabled him to survive from loans. He is now planning to keep the drumstick plants as long as he gets handsome income.

Another innovative farmer is **Venkateswarlu**. He is a poor resource farmer and who kept his land fallow this year (2018). With the help of SIF, he raised drumstick plants and earned Rs.35,000/- of profit with three harvests from his 2 acres of land of drumstick plantation under drip irrigation.

Settivari Srinu is yet another farmer from Giddalur of Prakasam district who has 1.5 acres land. He, with the aid of SIF has achieved success in drumstick cultivation and earned Rs.80,000/- of profit with 3 harvests from his 1.5 acres of drumstick.

Similarly, **Narayana Reddy** has about 2 acres of land and had tough time in 2018. He met SIF team and adopted the drumstick cultivation and as of now (January 2019) he started earning handsomely on his drumstick garden.

Such success stories in the cultivation of drumstick by SIF farmers are recorded in Andhra Pradesh (Anon, 2019).

Progressive farmers of Rampur and Dhanora villages of Adilabad district, Telangana State, abanded traditional and commercial crops and are now cultivating drumstick instead (Srinivas, 2017).

Ch. Om Prakash of Rampur village got a yield of 1.-1.5 quintals of drumsticks per day on his one acre plantation from January to May and sold for Rs.3000/- quintal during the period from January to June and earned Rs.3000 to 4,500/- per day (Srinivas, 2017).

Tamil Nadu

Tamil Nadu is the most important state in growing moringa for pods since ages. It is the pioneer state which did commandable research on moringa, developed and released improved and high yielding varieties for commercial cultivation, which crossed the boundaries of India. However, not many success stories are available from the state, except the following one.

Mr. Sadayandi (Plate A.1) has been cultivating moringa for the past 15 years in Pallapatti village of Dindigul district of Tamil Nadu State. He switched over from banana and sugarcane (water intensive crops) to moringa due to decline in water resources. He brought moringa seed from Valayapatti village, the major moringa growing village in the state and grew in his field. According to him, if properly maintained after planting in the field, moringa plants root deep into the soil and grow enduring drought; if they are pruned properly, they can yield for 50 years too, it can give good pods after 6 months of planting; but yields can be considerably high after 1.5 years. Initially he used chemical fertilizers and got very good yields for the first 3 years, but later the trees lost their leaves and flowers presumably due to excess use of chemical fertilizers and pesticides. Therefore, he switched over to organic farming of moringa. Since eleven years he is practicing organic farming only. He planted moringa in 8 acres. But, out of 8 acres of moringa he harvests pods from one acre only and from the rest he makes air-layers and sells them to enthusiastic farmers for raising plantation of moringa and earns Rs.21.0 lakhs from 70,000 air layers produced and sold @ Rs.30 per layer. He spent Rs.7.0 lakh and earned a profit of Rs.14.0 lakh from air-layers. He also sells drumsticks (pods) and leaves. By following his own cultivation methods he earns a profit of Rs.3,20,000/- per acre from pods (Stanley, 2019).

Fig. A.1 Success story.

There are many drumstick farmers in Tamil Nadu who are able to capitalize its market potential and earn handsome profits. One among them is **Mr. Aravindan** of Tuticorin district. While most of the moringa farmers use chemical inputs, in growing moringa, Aravindan stands alone by cultivating it on **Zero Budget** (Anon, 2017c). He used only chemical inputs at the beginning of his entry into farming, but 2 years later he switched over to **Zero budget** farming after reading a book **"PasumaiVikatan"** which inspired him. Since, then he dropped using chemical inputs and continued the use of "**Jeevamirthan**" which significantly reduced pest attack on moringa. From one hectare land, he could harvest about 60 t of pods by following **Zero budget** farming.

Table A.1 Expenditure and income for 1 ha of moringa as per Aravindan

S.No.	Item	Expenditure (Rs.)
1.	Ploughing the field	1,500
2.	Pit digging and planting	1,500
3.	Cost of stem cuttings	1,000
4.	Farm inputs	33,000
5.	Insecticides	3,000
6.	Labour for harvesting	10,000
7.	Transport to market	20,000
	Total	**70,000**
	Income through sale of 50t of moringa pods	4,00,000/-
	Net profit	3,30,000/-

At the same time, harvest has been increasing gradually by **Zero budget** farming and the pods are more robust with minimum wastage.

Gujarat

Oil and Natural Gas Commission (ONGC), is a public sector enterprise. It is has a NGO: Shroffs Foundation Trust. It is implementing a project "KalpaVriksha", which has objective of improving the health condition of tribal people of Chhota Udepur in Vadodara district of Gujarat State, by changing their food habits along with providing an additional source of income to improve the economic conditions of about 850 marginal and small farmers. The project promotes the cultivation of drumstick in 22 villages. KVK of Vadodara district under the project introduced the drought tolerant and high yielding variety of drumstick (PKM-2 of TNAU) in 2000-01. An estimated 1500 ha is presently under drumstick cultivation and it is spreading. Earlier farmers tried horticultural crops (ber and pomegranate) unsuccessfully and hence abanded them either due to pest attack or when they became non-profitable (Mohan, 2018) and are now cultivating moringa successfully.

Maharashtra

Of late, area under moringa cultivation in Maharashtra has increased, thanks to SIF. **Mr. Balasaheb Marale**, a school dropout farmer from Shaha village of Maharashtra has been cultivating drumstick with remarkable success in his drought prone village. Additionally he is helping other farmers in moringa cultivation. He is the author of a book on drumstick cultivation and has his own website on the subject. According to him, to hundreds of debt ridden farmers who committed suicides in Maharashtra and elsewhere, growing moringa is the solution. He opined that drumstick can be grown with rain water only without expensive irrigation technologies with good yields and income. Though the climate of Maharashtra is conducive for drumstick cultivation, the farmers avoided growing drumstick because of sentiment that it brings bad luck. However, Mr. Marale broke this sentiment by taking up drumstick cultivation successfully. He gave up his job as a machine operator in Pune in 1999 and took farming in his village. But he failed to get even medium yields in conventional crops. Instead of being discouraged with the conventional crops, boldly he took up cultivation of drumstick and became a role model for others. Before taking up drumstick farming he toured extensively both in Maharashtra and neighbouring states and met several drumstick growers, who, however, for want of adequate knowledge about moringa failed in their efforts to cultivate moringa. Next, he visited Tamil Nadu with a quest to have more information and knowledge about moringa cultivation and got enriched himself. By these tours he got convinced that drumstick indeed is the ideal crop for dry areas of Maharashtra. He returned to his village with seeds of 15 varieties of drumstick and a wealth of information and knowledge on drumstick and started its cultivation in his village successfully and profitably. He realized 440 to 1500 USD from an investment of 140 to 150 USD per acre of drumstick (Bose, 2015).

Mr. Santhosh Sambhaji Kalane of village Balwant Station, Taluka Sri Gonda, District Ahmednagar, has created a record in the production of drumstick by adopting the following techniques:

1. Land preparation in the month of May
2. Variety used was CO.1
3. Seed treatment with *Tricoderma viride*

4. Sowing seeds in June @ 2 seeds per pit

5. Spacing: 10′ x 10′

6. After 15 days of seeding 125 kg urea / ha was applied through ring method

7. 1 kg of organic manure per pit was applied after 15 days of seeding.

8. The same manure was applied many times at 45 days interval.

9. After two months of seeding 125 kg DAP, 125 kg NPK (10:26:26) were mixed and applied in a ring.

10. In January (next year) cow urine was applied through drip.

11. Applied a mixture of 250 kg urea + 125 kg NPK (10:26:26) + 125 kg DAP + 30 t FYM per ha and a plant growth regulator (name not indicated).

12. Needed inter culture was done, without irrigation during rainy season, but irrigation was given from December at an interval of 4 days through drip, and during the season irrigated through furrow method.

13. Branches were pruned after 15 days of seeding, as a result lateral branching increased

14. After a month of 1st pruning, medium height branches were pruned at a height of 6 feet.

15. Third pruning was done at a height of 9 feet to get profuse branching for fruit bearing.

16. All necessary plant protection measures were undertaken; stem borers, when located, were immediately destroyed with a peg and a mixture of kerosene and monocrotophos applied in the holes and closing with mud.

17. After a month of seeding, cypermethyne (5%) and chlorpyriphos mixture was prepared and used @ 2 ml / lit at fruiting.

18. A mixture of dichlorovos 76% EC (2 ml/lit) and thiophanate methyl 70% WP (0.5 g/lit) was sprayed.

19. To control powdery mildew disease microbutanyl 10 WP (0.5 g /lit) was sprayed.

20. A pheromone trap was installed for 20 plants each for the control of fruitfly.

The economics of first year cultivation is presented in the table – which indicates the huge profit reaped out of moringa cultivation. He earned more than Rs.2.0 lakhs in the first year in 6 months and is confident that the profits will increase in the second year onwards further (Staff, 2017).

Table A.2 Economics of drumstick cultivation (Rs/ha) during first year

Month	Yield (t)	Price (Rs/kg/ha)	Income (Rs.)
February	2.5	25	62,500
March	7.5	17	1,27,500
April	12.5	10	1,25,000
May	2.5	9	22,500
Total	25	-	3,37,500
Cost	-	-	65,000
Profit	-	-	2,72,500

According to Indian Meteorological Department, Maharashtra records, 55% deficit rainfall is received every year, hence the agriculture in the state is deemed to be rainfed one. Keeping the need for diversification of cropping pattern in mind, the farmers of dryland regions of Maharashtra have chosen moringa (*Moringa oleifera* Lam), considering its wide adaptability and drought tolerance. As per a study it was found that in moringa the cost: benefit ratio was 1:2.99 in the study area. Hence, with low cost of investment and high returns in the same year of cultivation, drumstick became an economically viable crop particularly in dry lands, which is amply demonstrated by the following success stories (Bose, 2019).

Prashant Jadhav, a farmer from Atpada village of Sangli district in Maharashtra grows moringa in his 3 acres, which gives him an assured yield of 15 t / acre. He markets his drumistcks in Pune, Kolhapur and Sangli @ Rs.20-25 /kg. He follows sustainable production practices in moringa farming like drip irrigation and grows a moringa variety (name not quoted) which fruits nearly 10 months (January to October) in a year. He learnt about moringa and its farming from "You Tube".

Shantanu Chandrasekhar Mahale, a Telecom Engineer from Nagpur University of Lohi village in Yavatmal, runs a hardware solution firm. For a decade his family has been growing soybean and redgram (pigeon pea) without good returns, but ever since he took over the farming and started moringa plantation on the family's 20 acre land in 2017, their fortune has changed for better. He grows moringa in arid zone like Yavatmal with assured income and makes moringa leaf powder too. He is popularizing moringa cultivation through his website. He learnt much about moringa through internet.

Shyamsunder Jaygude of Kelawade in Pune has 240 moringa trees planted on the edges of his 20 acre farm. A pioneer farmer, he introduced Puneites to exotic vegetables like broccoli, Thai chillies, Bell pepper etc. He adopted organic methods of cultivation since a decade. He sells drumsticks (pods), leaves and leaf powder. He sells pods at a fixed rate of Rs.80/kg. He makes approximately Rs.5.4 lakh annually from sale of pods, leaves and leaf powder.

BalaShivaji Patil of UplaiKhurd village in Madha taluk of Solapur district is popularly known as **"Lakhpati Shetkar"** (farmer). Moringa has not only brought riches to his family but also fame. He grows moringa on 4 acre land, each acre fetching around Rs.4.0 lakh. He authored a book entitled **"Crorepati Benvel Shewga"** (meaning: Moringa can make you a crorepati), published in 2014 and sold over 10,000 copies so far.

A farmer, **Appa Karmakar's** is another successful story of growing moringa in Angon village with a sparse vegetable land scape, a temperature of 40^0C and squat houses. His 3 acre drumstick farm has become a shining example of dry land farming. He is a post graduate, started his drumstick farming in the year 2012, with intercrops of chillies, papaya, pomegranate and guava, besides marigold, the traditional crop for control of pests. His farm produces 50 t of drumsticks every year, which sells for Rs.30 to 80 per kg (depending on season) in the whole sale markets fetching approximately Rs.6.0 lakhs per year (Mohan, 2018).

Such successful farmers of moringa cultivators of Solapur, who have seen prosperity, have become common in the drought prone areas of Maharashtra and are being emulated by others in other dry zones of Maharashtra. Further, farmers of dry land regions of Maharashtra have realized that "Moringa" is a climate resilient crop, which requires little effort and investment to grow, but fetches handsome prices and profits (Bose, 2019).

References

Abdel Wanise, F.M., Saleh, S.A., Ezzo, M.I., Helmy, S.S. and Abodahab, M.A. 2017, Response of moringa plants to foliar application of nitrogen and potassium fertilizers. Acta Horticulturae. 1158: 187-193.

Abdul Das, 2007. Economic importance of *Moringa oleifera* Tafa Local Government Area of Niger State NDF Project. Federal College of Forestry Mechanization, Kaduna, Nigeria.

Adbul Kareem, A., Sadakathulle, S. and Subramaniam, T.R. 1974. Note on the severe damage of moringa fruits by the fly *Gitona* spp. (Drosophillidae: Diptera). South Indian Horticulture.22: 77.

Adebayo, A.G., Akintoye, H.A., Shokulu, A.O and Olatunji, M.T. 2017. Soil chemical properties and growth response of *Moringa oleifera* to different sources and rates of organic and NPK fertilisers. International Journal of Recycling of Organic Waste in Agriculture. 6 (4): 281-287. https://doi.org/1007/S40093-017-0175-5.

Agyenin – Boateng, S., Zickermann, J. and Kovnaherms, M. 2006. Poultry manure effects on growth and yield of maize. West African Journal of Applied Ecology. 9:1-11.

Ali, M.S., Azam, F. and Chaturvedi, O.P. 2007. Occurrence of host spectrum, host specificity of bark eating caterpillar, *Indarbela quadrinotata* Walk in relation to trees of Bihar. Tropical Forest. 23 (3/4): 59-64.

Allison, F.E. 1973. Soil organic matter and its role in crop production. Amsterdamk. Scientific Publishers Co. Pp. 16-23.

Amaglo, N.K., Timpo, G.M., Ellis, W.O. and Bennette, R.N. 2007, Effect of spacing and harvest frequency on the growth and leaf yield of moringa (*Moringa oleifera* Lam), a leafy vegetable crop. Proceedings "Moringa and other highly nutritious plant resources: Strategies, Standards and Markets for a better impact on nutrition in Africa", Accra, Ghana.

Amanullah, M.M., Sekar, S. and Muthukrishnan, P. 2010. Prospects and potential of poultry manure. Asian Journal of Plant Science. 9: 172-182.

Anitha.V., Balaji Naik, R. and Sudhakar Reddy, R. 2013.Profitable moringa cultivation. *Rythunestham*. March 2013: 25-26.

Anjaneyamurthy, J.N. 1985. Studies on seasonal incidence and insecticidal control of fruitfly *Gitona* sp. leaf eating caterpillar, *Noorda blitealis* Walk and legume aphid *Aphis croccivora* Koch on annual moringa. M.Sc.(Ag.) Thesis. Tamil Nadu Agricultural University, Coimbatore, Tamil Nadu.

Anjaneyamurthy, J.N. and Regupathy, A. 1988.Studies on seasonal incidence of fruit fly, *Gitona* sp. Proceedings of workshop on Pest surveillance and forecasting, September 16-17, Tamil Nadu Agricultural University, Coimbatore, India.

Anjaneya Murthy, J.N. and Regupathy, A. 1989.Insecticidal control of fruit fly *Gitona* sp., leaf eating caterpillar *Noorda blitealis* Walk and the aphid, *Aphis craccivora* Koch on annual moringa.South Indian Horticulture. 37 (2): 84-93.

Anjaneyamurthy, J.N. and Regupathy, A. 1992.Seasonal incidence of moringa fruit fly, *Gitona* sp. South Indian Horticulture. 40 (1): 43-48.

Anjulo, A., 2009. Screening Moringa accessions for resistance to moringa moth, *Noorda blitealis* Walker (Crambidae: Noordidnae). Indian Journal of Forestry. 32 (2): 243-250.

Anonymous, 2002.*Moringa oleifera* a multipurpose tree – The Tropical Advisory Service, HDRA.The Organic Organization, Ryton Organic Gardens, Coventry, UK, HDRA, Publishing. Pp.18

Anonymous, 2005. Annual Report 2004-05. Department of Vegetable Crops, Horticulture College and Research Institute, Tamil Nadu Agricultural University, Periakulum.

Anonymous, 2006.14 "Moringa" National Research Council 2006. Lost crops of Africa: Volume II Vegetables, Washington D.C. The National Academic Press, Page 251. doi:10.17226/11763.

Anonymous, 2010a.Significance of Bio-fertilizers in Crop Production. Plant Horti-Tech. 10 (4): 44-47.

Anonymous, 2010b.Malunggay Propagation Technique. Seek No. More Entrepreneurs Window of Opportunities. Agricultural Official Endorses Protocol for Moringa Mass Propagation. The Philippine Star, Agriculture 14 January 2010.

Anonymous, 2012a.Agriculture Forum.All about moringa. green agrow. blog spot. In/2012/11/all-about-moringa-html.

Anonymous, 2012b. Drumstick cultivation practices. www.crops/aaqua. persistant.com/aaqua/torun/view thread? Thread = 23593.

Anonymous, 2012c.Moringa-An Excellent Source of Nutrition. Technical Bulletin # 24.USAID. pdf.usaid.gov/pdf-docs/pnaeb 605 pdf.

Anonymous, 2013. Flowering and flowering seasons in moringa. www.farm nest.com/forum/agriculture-crop-flowering and flowering seasons in moringa.

Anonymous, 2014a. Moringa in Ancient Egypt.www.facebook.com>posts> moringa-in-ancient-egypt.

Anonymous, 2014b. Spider mites love moringas. www.moringaceae.org/1/ pests / 2014/01/spider-mites-love-moringa.

Anonymous, 2015a. *Moringa oleifera.* http://en.wikipedia.org./wiki/Moringa-oliefera.

Anonymous, 2015b. Capitalizing on "Drumstick cultivation". Promoting Drumstick cultivation to Meet Changes of dryland farming and climate change. Training Model. Institute of Horticulture Technology. 2 (7):

Anonymous, 2015c. Moringa Facts, History and Legend.www.its moringa.com/ 1/history-of-moringa.

Anonymous 2015d. TNAU, Pests of Moringa: Major pests, pod fly, TNAU Agritech Portal Crop Protection, Tamil Nadu Agricultural University, Coimbatore, India. http://agritech.tnau.ac.in/crop-protection/moringa /moringa-4.

Anonymous, 2017. Zero Budget drumstick… profit of 3 lakhs annually..! www.vikatan.com>additives>miscellaneous.

Anonymous, 2019. Moringa in Ancient Egypt. www.sinai-moringa.com/moringa in ancient Egypt.

Antwi – Boasia Ko, C and Enninful, R.2011. Effects of growth medium, a hormone, and stem-cuttings maturity and length on sprouting in *Moringa oleifera* Lam. Journal of Horticultural Science & Biotechnology.86 (6):619-625.

Anwar, F., Latif, S., Ashraf, M. and Gilani, A.H., 2007. *Moringa oleifera*: A food plant with multiple medicinal uses. Phytotherapy Research. 21: 17-25.

Armella de Saint Sauveur and Broin, M. 2002. Growing and processing of moringa leaves. Moringa and Plant Resources Network 211 rue du Faubourg Saint Antoine 75011 Paris, France. www.moringanews.org.

Arun, P.R., Anand Prem, R. and Sarita, S. 2011. Comparative analysis of preservation technique on *Moringa oleifera*. Asian Journal of Food and Agro-Industry. 4(2): 65-80.

Atwal, A.S.1986. Agricultural Pests of India and South-East Asia. Kalyani Publishers, New Delhi Pp.215-216.

Auwalu, B.M. 2009. Agro-economic management practices for commercial production of moringa. A paper presented at a two day workshop on

Sustainable production and Commercialization of Moringa, organized by Kano State Government, Nigeria.

Ayyar, T.V.R. 1963. Pests of drumstick. Handbook of Economic Entomology for South India, Madras.Pp.249-250.

Baba.M.F., Yakubu, G., Yelwa, J.M. and Haruna, L. 2015.Costs and Returns of Moringa (*Moringa oleifera*) Production in Zuru Local Government Area Kebbi State, Nigeria. New Yark Science Journal. 8(1): 36-40. http://www.siencepub.net/newyork.6.

Banu, K and Channabasavanna, G.D. 1972. Plant feeding mites of India. A preliminary account of the spider mite *Tetranychus neocaledinicus* (Andre) (Acarina: Tetranychidae). Mysore Journal of Agricultural Sciences. 6(3): 253-268.

Basu Choudhury, J.C. and Mishra, M.P. 1977. Immature stages of Indian Coleoptera (31). Description of egg, larva and pupa of cashew stem and root borer, *Plocaederus ferrugineus* (L), (Coleoptera: Cerambicidae). Indian Forest Records (New Series). Entomology. 12 (1): 1-5.

Beaulah, A 2002. Growth and development of moringa (*Moringa oleifera*) cv. PKM-1 under organic and inorganic systems of production. Ph.D. Thesis, Tamil Nadu Agricultural University, Coimbatore, India.

Beaulah, A., Swaminathan, V., Rajangam, J. and Ponnuswami, V. 2010. Moringa – A nature's gift. Horticultural College and Research Institute, Tamil Nadu Agricultural University, Periakulam, Tamil Nadu, India.

Beaulah, A., Vadivel, E. and Rajadurai, K.R. 2004.Effect of organic and inorganic fertilizers on growth characters of Moringa (*Moringa oleifera* Lam.) cv. PKM-1.South Indian Horticulture. 52 (1-6): 183-193.

Benerji, B.K. and Batra, A. 2012. Moringa – Poor Man's Tree with Rich Nutritive and Medicinal value. Indian Farmers' Digest. 45 (1): 15-16.

Benthal, A.P. 1946. Trees of Calcutta and its neighbor hood. Thacker Spink & Co (1933) Ltd., Calcutta.

Bezerra, A.M.E. Medeiros Filho, S., Freitas, J.B.S and Teofilo, E.M. 2004. Evaluation of the quality of drumstick seeds during the storage.Ciencia e Agrotechnologia. 28 (6): 1240-1246.

Bharathi, C.S., and Pugalendhi, L. 2014. Canopy management and chemical manipulation on pod characteristics of annual moringa (*Moringa oleifera* Lam.) cv. PKM-1 during off season production.Trends in Biosciences. 7, (15): 1886-1890.

Bose, H.K. 2015. Farmer's drumstick beats drought. www.news.bbc.co.vk>southAsia/Farmer's-drumstick-beats-drought.

Bose H.K. 2019. These Maharastra farmers are earning lakhs by growing a desi super food. www.thebetterindia.com>Stories

Brunda Kumari, M.S. 2014. Studies on the insect defoliators of drumstick with special reference to *Noorda blitealis* Walker (Lepidoptera: Crambidae). M.Sc. (Hort.) Thesis, University of Horticultural Sciences, Bagalkot, Karnataka.

Burkill, J.H. 1966. A dictionary of economic products of the Malay Peninsula, Vol.2.Art Printing Works Publishers, Kaulalumpur, Malaysia.

Butani, D.K. and Jotwani, M.G. 1984. Insects in vegetables. Periodical Expert Book Agency, Delhi, India.Pp. 356.

Butani, D.K. and Verma, S. 1981. Insect pests of vegetables and their control – drumstick.Pesticides. 15: 29-31.

CABI, 2016. Data sheet report for *Batocera rubus*. (http://www.cabi.org /epe/datasheetreport/dsid=8572).

Caceres, A., Freire, V., Gifron, L.M, Aviles, O. and Pacheco, G. 1991. *Moringa oleifera* (Moringaceae): ethno botanical studies in Guetamela. Economic Botany. 45 (4): 522-525.

Chaves, L.H.G., Viegas, R.A., Vasconcelos, A.C.F. de and Vieira, H. 2005.Effect of potassium on moringa plant growth in Nutrient solution Revista de Biologia E ciencia Da Tetra.5 (2).

Cherian, M.C. and Basheer, M. 1938.A new cecidomyiid pest of moringa. Madras Agricultural Journal. 26 (3): 92-95.

Chinnadurai, M., Shivakumar, K.M. and Kanaka, S. 2010. Marketing strategies for moringa. In: L. Pugalendhi, B. Singh, S. Natarajan and S. Makesh (eds). Moringa – A compendium of brain storming session on moringa 23rd September 2010 held at TNAU, Coimbatore, Tamil Nadu, India by Dept. of Vegetable crops., TNAU, Coimbatore & AICRP on Vegetable crops, IIVR, Varanasi, India. pp. 151-158.

Chowdappa, P. and Krishna Kumar, N.K. 2010. Diseases and pest management in moringa. In: Moringa.-A compendium of brainstorming session on moringa. (eds). L.Pugalendhi, B.-Singh, Natarajan, S and Makesh, S 2010). Organized by Dept. of Vegetable crops. TNAU, Coimbatore and AICRP on Vegetable Crops, IIVR, Varanasi, India, on 23rd September 2010.

Copeland, L.O. and McDonald, M.B. 1995. Seed viability testing. In: Principles of Seed Science and Technology (3rd Edn). Pp. 111-126. Chapman and Hall, New York.

Damodaran, T. 1998. Organic and inorganic nutrients and packing methods to prolong the shelflife of moringa. M.Sc.(Hort) Thesis, Tamil Nadu Agricultural University, Coimbatore, Tamil Nadu, India.

Damodaran, T., Anbu, S., Azhakiamanavalan, R.S. and Vennila, P. 1999. Packaging methods to prolong the shelf life of moringa (*Moringa oleifera* Lam) cv. PKM-1 during transit. South Indian Horticulture. 47 (1-6): 292-294.

Dania, S.O., Akpansubi, P and Eghagara, O.O. 2014. Comparative Effects of Different Fertilizer Sources on the Growth and Nutrient content of moringa (*Moringa oleifera*) seedling in a greenhouse trial. Advance in Agriculture 2014 (2014), Article ID 726313, 6 pages. https:dxdoi.org.10.115/2014/726313.

Dao, M.C.E., Traore, M.P., Souleymane, O., Relphine, B., Oueddrago, S. 2015. Ravageurs de planches maraich resde *Moringa oleifera* dams la region due centre (Burkina Faso). Journal of Animal and Plant Sciences. 25 (2): 3857-3869.

David, B.V. 1961. A Diaspidine scale on moringa in South India. Madras Agricultural Journal. 48 (6): 227.

David, B.V. 2001.Elements of Economic Entomology. Popular Book Depot.Chennai.562 pp.

David, B.V. and Kumaraswami. T. 1982. Pests of moringa. Elements of Economic Entomology' Popular Book Depot, Madras. pp. 122-124.

David., B.V. and Ramamurthy, V.V. 2016. Elements of Economic Entomology. Eight edition.Brillion Publishing, New York, New Delhi. p. 398 and 147-148.

David, K.S. 1958. Notes on South Indian Aphids. III – Lachinae to Aphidimae (Part) Indian Journal of Entomology.1(3): 77.

Davoodi, M.G., Vijayanand, P., Kulkarni, S.G., and Ramana, K.V.R. 2007.Effect of different pre-treatments and dehydration methods on quality characteristics and storage stability of tomato powder. Food Science and Technology. 40: 1832-1840.

Demuclenaere, E. 2001. *Moringa stenopetala*, a subsistence resource in the Konso district. Proceedings International Workshop on Development Potential for Moringa products. Dar-ES-Salam, Tanzania.2-29.

Devi,C. 2003. Studies on the influence of pruning and growth regulators on annual moringa (*Moringa oleifera*) cv. PKM-2. M.Sc.(Hort) thesis. Tamil Nadu Agricultural University, Coimbatore, Tamil Nadu.

Devi, C., Vijaya Kumar, R.M. and Nainar, P. 2007. Studies on influence of spacing and pinching treatments for pod characteristic and quality attributes of annual moringa (*Moringa pterygospema* Gaertn) cv. PKM-2.South Indian Horticulture. 55 (1-6): 84-91.

Dhakar, R.C., Maurya, S.D., Pooniya, B.K. Bairwa N., Gupta, M and Sanwarml. 2011. Moringa: The herbal gold to combat malnutrition. Chron Young Science, 2: 119-125. http://www.cysonline.org/text.asp?2011 /2/3/119/90887. DOI: 10.4103/2229-5186.90887.

Doerr, E and Williams, N. 2007-09. Moringa Manual 2008. Petanzama.org/repository/environment/.../moringa o/o20 manul / moringa manual 2008.

Du, L.V. and Tuong, T.P. 2002.Enhancing the performance of dry seeded rice: Effects of Seed priming, seedling rate and time of seedling. In: Direct Seeding Research Strategies and Opportunities (Eds: Pandey M. Mortimer, L. Wade, T.P. Tuong, K. Lopes and B. Hardy). International Rice Research Institute, Manila, Philippines, Pp. 241-246.

Duke, J. 1978. The quest for tolerant germplasm, P.1-61. In: G.Young (ed) Crop tolerance to suboptimal soil conditions. American Society Agronomy. Spec. Symposium 32. Madison Wiscousin, USA.

Erin, H. 2007. "organic Farming" Microsoft student 2008 (DVD) WA: Microsoft Corporate 2007. Microsoft Encarta, 2008(c): 1993-2007. Microsoft Corporation.

Exodus.Old Testment.15-25.

Fahey, J.W. 2005.*Moringa oleifera*: A review of the medical evidence for its nutritional, therapeutic and prophylactic properties. Trees of life Journal 1 (5): 5-11. http://www.tfljournal.org/articles.php/20051201124931586.

Fallik, E., Aharoni, Y., Grinberg, S., Copel, A. and Klein, J.D. 1994. Post harvest hydrogen peroxide treatment inhibits decay in egg plant and sweet red pepper. Crop Production. 13: 451-454.

Farooq, M., Basra, S.M.A., Hafeez, K. and Ahmad N. 2005.Thermal handling: a new seed vigor enhancement tool in rice. Integrated Plant Biology. 47: 187-193.

Farooq, M., Basra, S.M.A., and Rehman, H. 2006. Seed priming enhances emergence, yield and quality of direct seeded rice. International Rice Research Notes. 31: 42-44.

Farooq, M., Basra, S.M.A., Rehman, H. and Hussain, M. 2008. Seed Priming with polyamines improves the germination and early seedlings growth in fine rice. Journal of New Seeds. 9: 145-155.

Farse, J.G., Sontakke, P.M., Damodhar, V.P., Pawar, P.M. and Nawghare, P.D. 2006.Effect of severity of prunning on growth, flowering duration and yield in drumstick (*Moringa pterygosperma* Gaertn).Journal of Asian Horticulture. 2 (3): 215-217.

Farse, J.G., Sontakke, P.M., Pawar, P.M., Nawghase, P.D. and Damodhar, V.P. 2004.Pod number, edible portion and quality in drumstick (*Moringa pterygosperma* Gaertn) as influenced by severity of pruning. Karnataka Journal of Horticulture. 1(1): 89-92.

Fasiha, M. and Srivastava, R.P. 1988.Natural occurrence of *Beauweria bassiana*, an entomogenous fungus on bark eating caterpillar, *Indanbela* spp. Indian Journal of Plant Pathology. 6(1): 11-16.

Fletcher, T.B. 1919. Annotetred list of Indian Crop pests. Proceedings of the 3rd Entomological Meeting. Pusa, p. 102-103.

Foidl, N., Makkar, H.D.S. and Becker, K. 2001. The potential of *Moringa oleifera* for agricultural and industrial uses. In: The Miracle tree / the multipurpose attributes of moringa (Ed. Lowell. J. Fuglie) CTA.USA.

Fuglie, L J 1999. The miracle tree – *Moringa oleifera*: Natural Nutrition for the tropics. Church World Service, Dakar, Senegal: 18.

Fuglie, L.J. 2001. The Miracle tree: *Moringa oleifera*: Natural Nutritional for the tropics. Training Mannal.Church World Service, Dakar, Senegal.www.moringatrees.org/moringa/miracletree.html>.

Fuglie, L.J. and Sreeja, K.V. 1998. Cultivation of Moringa: Growing Moringa, Part II (F. Annenberg. Ed.) Wonder Plants of the world. The Morning Tree, pp.6. http://moringaforms.com/cultivation-of-moringa/

Fugro, P.A., Rajput, J.C. and Parulekar, Y.R. 1996. Fungicidal control of Fusarium wilt of drumstick seedlings. Pestology. 22 (1): 5-6.

Ghori, T.K. and Anusuya, B. 2017. Responses of *Moringa oleifera* Lam (Drumstick) plant on inoculation with *Frateuria aurentia* (Potassium mobilizer) and plant growth promoting *Rhizomicro*-organisms (PGPR). International Journal of Scientific and Engineering Research. 8 (4): 889-891.

Gilletti, M.R.T. 1997. Brief notes on some species of micro-moths newly recorded from Al Ain: (Lepidoptera: Micro-Heterocera: Pyraldae). Tribulus. 7(1): 19-20.

Giraldo, L.F., Forero, R.A., Salazar, C.R. and Torres, R. 1977. The effect of packaging and potassium permanaganate on the storage of tomatoes under room conditions. Horticulture Abstracts. 12 (4): 393-405.

Gomati et al., 1999. Quoted by Pugalendhi et al., 2010.Moringa A compendium of brain storming session on moringa. Organized by Department of Vegetable crops. TNAU, Coimbatore and AICRP on Vegetable Crops, IIVR, Varanasi, India, 23rd September, 2010 at Coimbatore.

Gupta N. 2018. Need guidelines on drumstick cultivation in Valsad, Gujarat. Agricultural and Industry Survey. 28 (4): 35.

Halder, J and Rai, A.B. 2014. New record of leaf caterpillar, *Noorda blitealis* Walker (Lepidoptera: Pyralidae) as fruit and seed borer of drumstick, *Moringa oleifera* Lam. Journal of Plant Protection and Environment. 11 (2): 6-9.

Hanchinamani, C.N. and Madalgeri, B.B. 1994. Response of drumstick cultivars to different nutrient levels. Karnataka Journal of Agricultural Sciences. 7 (1): 18-22.

Haou Vang, L.C. Nagakou, A., Yemefack, M. and Mbailao, M. 2017.Growth responses of *Moringa oleifera* Lam. as affected by various amounts of compost under greenhouse conditions. AOAS 62: 221-226. https://doc.org/10.1016/j.aoas.2017/12.004 cross Ref. Google Scholar.

Harris, D., Tripathi, R.S. and Joshi, A. 2002. On farm seed priming to improve crop establishment and yield in dry direct seeded In: Direct Seeding: Research Strategies and Opportunities. (Eds. S. Pandey; M.Mortimer; L. Wade; T.P. Toung; K. Lopes and B. Hardly) International Rice Research Institute, Manila, Philippines. Pp. 231-240.

Harsh, N.S.K. and Ojah, B.M. 2002.A biocontrol formulation for post-emergence damping off of *Moringa pterygosperma*. Biochemistry of Microbes and Sustainable Utilization: 108-115.

Hazarika, B.N. and Ansari, S. 2007. Biofertilizers in fruit crops: A review Agricultural Reviews, Agricultural Research Communication Centre, India.

Honnalingappa, Y.B. 2001. Insect pests of drumstick (*Moringa oleifera* Lam) with special reference to the bio ecology and management of leaf eating caterpillar, *Noora blitealis* Walker (Lepidoptera: Pyralidae). M.Sc.(Ag.) Thesis. University of Agricultural Sciences, Bangalore, Karnataka.

HDRA. 2002. *Moringa oleifera*: a multipurpose tree. http://www.gardenorganic.org.uk/pdfs/international-progamme/ moringa. > HDRA-the organic organization. Coventry. U.K.

Hsiao, C. and Lauchli, A. 1986. Role of potassium in plant – water relation. In: Advances in Plant Nutrition, 2 Tinker and Lauchli, (Eds) New York: Praegers P. 281-312.

Hurly R.F., Staden, J.V. and Smith, M.T. 1991. Improved germination in seeds of guayule (*Parthenium argentatum* Gray) following polyethylene glycol and gibberllic acid pre-treatment. Annals of Applied Biology. 118: 175-184.

IIHR. 2002. Quoted by V. Ponnuswami, 2012. In: Advances in Production of Moringa. All India Coordinated Research Project – Vegetables crops. Horticultural College and Research Institute, Tamil Nadu Agricultural University, Periakulam, Tamil Nadu, India.Pp 20.

Jadav, R.G., Patel, H.C., Masu, M.M., Parmar, A.B., Sitapara, H.H. and Patel, H.D. 2009.Effect of spacing and severity of pruning on yield of drumstick cv. PKM-1.Asian Sciences. 4 (1&2): 4-9.

Jadav, R.G., Patel, H.C., Masu, M.M., Sitapara, H.H., Parmar, A.B. and Patel, H.D. 2010.Fertilizer requirements of drumstick cv. PKM-1.International Journal of Agricultural Sciences. 6(1): 220-225.

Jahan, S.I. and Khatun, MAAR. 2005. *In vitro* regeneration and multiplication of year-round, fruit-bearing *Moringa oleifera* L. Journal of Biological Sciences. 5 (2): 145-148.

Jahn, S.A.A. 1991. The traditional domestication of a multipurpose tree *Moringa stenopetala* (Bak.f) Cuf.In the Ethiopia Rift Valley. AMBIO 20: 244-247.

Jahn, S.A.A. 1996. On the introduction of a tropical multipurpose tree to China:Traditional and potential cultivation of *Moringa oleifera* Lamark. Senckenbergiana Biologica. 75 (1-2): 243-254.

Jahn, S.A.A. Mushad, H.A. and Burgstaller, H. 1986. The tree that purifiers water. Cultivating multipurpose Moringaceae in the Sudan. Unasylva, 38: 23-28.

Jamaludin, M.H. 2006.Food preservation using solar drying technology. Journal of Agriculture Science. 21: 12-14.

Janakiram, T., Armugam, T., Tamil Selvi, N.A. and Premalakshmi, V. 2017.Hort Fact sheet. Moringa at a glance. Indian Horticulture.62 (4). Cover III.

Jaskani, M.J., Abbas, H., Khan, M.M., Qasim, M., Sultana,R. and Khan I.A. 2008. Effect of growth hormons on micro propagation of *vitis vinifera* L. cv. Perlett.Pakistan Journal of Botany. 40: 105-109.

Joseph, S. 2007. Chopter 18.Drumstick. In: K.V. Peter(ed) Under utilized and Under exploited horticulture crops. Vol.2. pp. 311-321. New India Publishing Agency, New Delhi, India.

Joshi, R.C., David, B.V. and Kant, R. 2016. A review of the insect and mite pests of *Moringa oleifera* Lam. Agriculture for Development, 2016 No. 29. Pp. 29-33.

Justice, O.L. and Buss, L.N. 1978. Principles and Practices of seed storage. Agriculture Hand Book No.506.pp. 289.USDA, Washington DC, USA.

Kalalbandi, B.M., Ziauddin, S. and Solanke, S.P. 2014. Effect of number of branches on morphological characters and yield of drumstick (*Moringa pterygosperma* Gaertn). Agriculture Sciences Digest. 34 (1): 73-75.

Kadam, V.D., Giri D.G., Solunke, P.S., Deshmukh, M.R. and Giri, M.D. 2006.Effect of integrated nutrient management on growth and yield of drumstick varieties (*Moringa oleifera*).Crop Protection and Production. 2(2): 22-24.

Kader, M.M. and Shanmugavelu, K.G. 1982. Studies on performance of annual drumstick (*Moringa pterygosperuma* Gaertn) at Coimbatore. South Indian Horticulture. 30: 95-98.

Kant, R.C. and Joshi, R.C. 2017.Survey of insect pests of *Moringa oleifera* in Samoa. Acta Horticulturae. 1158: 195-199.

Kantharajah, A.S. and Dodd, W.A. 1991. Rapid clonal propagation of *Moringa oleifera* Lam. using tissue culture. South Indian Horticulture. 39(4): 224-228.

Kanthaswamy, V. 2006.Studies on phenology and floral biology in *Moringa oleifera* Lam. International Journal of Agricultural Sciences. 2 (2): 341-343.

Kashyap, A.S., Bharatiya, A. and Kumar, S. 2009. Moringa to supply food, fodder and drug. Indian Horticulture. 54 (3): 51 &54.

Kathiresan, G., Manoharan, M.L. and Velusamy, 1999.Package of practices for drumstick PKM-1.Indian Horticulture. 44(2): 32.

Kaur, R and Kaur, S. 2018. Biological alternates to synthetic fertilizers: efficiency and future scopes. Indian Journal of Agriculture Research. 52(6): 587-595.

Khanna, K.K. and Chandra, S. 1977.Some new leaf spot diseases. Proceedings of National Academy of Science, India. 47 (3): 118-124.

Kishchuk.B.E. 2000. Calcareous soils, their properties and potential limitation to conifer growth in Southern British Columbia and Western Alberte: a literature review. International Research NOR-X-370 (British Columbia: Ministry for Invermere).

Kotikal, Y.K. and Mahesh Math. 2016. Insect and non-insect pests associated with drumstick, *Moringa oleifera* (Lamk). Journal of Global Biosciences. 5(4): 3902-3916.

Kshirsagar, C.R. and D'Souza, T.F. 1989. A new disease of drumstick. Journal of Maharashtra Agricultural Universities. 14 (2): 241-242.

Kumar, A and Yadav, V.P. 2007. Vesicular Arbuscular Mycorphyzae: a Catylyst for tree growth. Agriculture Update. 2(1): 9-12.

Kumar, N. and Pugalendhi, L. 2010. Intensive moringa cultivation for food, leaf and oil yield pp. 119-133. In: L.Pugalendhi, Singh, B., Natarajan, S., Makesh, S (Eds). Moringa. A compendium of brain storming session on moringa. 23rd September 2010. Coimbatore, TNAU, Dept. of Vegetable Crops, TNAU, Coimbatore and AICRP on Vegetable crops. IIVR, Varanasi.

Kumar, S., Singh, R. and Saini, D.C. 2013.*Moringa oleifera*, a new host record of *Cercospora apii* S. lat.From Uttar Pradesh, India. Plant Pathology and Quarantine, 3(1): 12-13. Doi10.5913.ppq/3/1/2

Kumari, M.S.B., Kotikal, Y.K. Narabenchi, G. and Nadal, A.M. 2015. Bioefficacy of insecticides, botanicals and bio-pesticides against the leaf eating caterpillar, _Noorda blitedis_ Walkar on drumstick. Karnataka Journal of Agricultural Science. 28 (2): 193-196.

Lalitha Kameshwari, P. 2004. Profitable cultivation of PKM-1 moringa. Annadata. May 2004: 8-9.

Lezcano, J.C., Alonso, O., Marialys, T. and Martizez, E. 2014. Fungal agents associated to disease symptoms in seedlings of _Moringa oleifera_ Lam Pastos Y Forrajes. 37 (2): 228-232.

Litsinger, J.A. 2014. Pesticide evaluation and safe use practices for USAID and Catholic Relief Services (CRS) Project in Niger on Development Food Aid Programme (DFAP) 14 January – 28 February 2014. Niger. 148.

Little, E.L. and Wadsworth, F.H. 1964.Common trees of Puerto Rico and the Virgin Islands. Agricultural Handbook. 249. U.S. Dept. of Agriculture, Washington D.C. USA.

Madalageri, B.B., Ganiger, V.M., Bhuvaneswari, G. and Sneha Shattar 2010.Research and Development activities of drumstick at University of Horticultural Sciences, Bagalkot. In: L. Pugalendhi, Singh, B., Natarajan, S. Makesh, S. 2010. (eds) Moringa. A compendium of brainstorming session on moringa, 23rd September 2010, TNAU, Coimbatore, Dept. of Vegetable crops, TNAU, Coimbatore and AICRP on Vegetable crops, IIVR, Varanasi. Pp. 100-105.

Madhumathi, V.M., Srivastava, P, Mohan, P.S. and Purohit, V.M. 1995. _In vitro_ micro propogation of _Moringa pterygosperma_. Phytomorphology, 45 (3&4): 65-69.

Madhusudan Reddy, D. 1999. Better Management practices in moringa cultivation. Annadata, January 1999: 40-42 & 49.

Madhusudan Reddy D. 2002.Insect pests and diseases in moringa and their control. Annadata, August 2002: 25-26.

Mahesh Math and Kotikal, Y.K. 2014. Studies on insect pests of drumstick, _Moringa oleifera_ Lamk. Indian Journal of Plant Protection. 42 (4): 461-464.

Mahesh Math and Kotikal, Y.K. 2015. Seasonal incidence of insect pests, natural enemies and pollinators in drumstick (_Moringa oleifera_ Lamk) ecosystem. Green Farming. 6 (17): 144-148.

Mahesh Math, Kotikal, Y.K. and Gangadhara, N. 2014. Management of drumstick pod fly _Gitona distigma_ (Meigen). International Journal of Advances in Pharmacy, Biology and Chemistry. 3 (1): 54-59.

Majhi, S. 2013. Nutritional value of *Moringa oleifera* as a dietary supplement. Master of Pharmacy. Thesis, Jadavpur University, Kolkata, India.

Malik, S.K. 1992. Multipurpose tree species. Indian Farmers' Digest. 25 (9-10): 29-32.

Mandokhot, A.M., Fugro, P.A. and Gonkhalekar, S.B. 1994.A new disease in *Moringa oleifera* in India. Indian Phytopathology.17 (4): 443.

Manh, L.H., Dung, N.N.X and Xuana, V.T. 2003.Biomass production of some legumes in the hilly areas of Thinh Bien District, An Giang Province. In: Proceedings of Final National Seminar Workshop on Sustainable Livestock Production on Local Feed Resource (Eds: Reg Prestan and Brain Ogle) HUAF - SA REC, HUC city, 25-28 March 2003: http://www.mekarn.org. /sarec O3/manhcantho2.htm.

Marie, J. Which zones to grow *Moringa oleifera*. In Home guide Atgate.com/zones-growing-moringa oliefera-65509-html

Martin, F.W. and Rubertie, R.M. 1975. Edible leaves of the Tropics, U.S. Agency for International Development and Agricultural Research Service, US Dept. of Agriculture, Mayaguez, Puerto Rico.

Marschner,H. 1995. Mineral nutrition of higher plants. 4[th]Printing (1999) (London: Academic Press) pp. 889.

Mathew S.K. and Rajamony, L. 2004.Flowering biology and palynology in drumstick (*Moringa oleifera* Lam.).Planter. 50: 357-368.

Mohan, C. 2018. Drumstick can make you Lakhpati: Attraction for the benefit of the farmers. Krishijagaran.com>success story>drumstick-can-make-you-lakpati-attraction-for-the-benefit-of-the-farmers.

Mohan, S., Gopalan, M. and Sreenarayanan, V.V. 1993.Fish meal waste as an attractant for economically important flies of agricultural crops. Bioresources Technology. 43(2): 175-176.

Mohan, V., Purohit, M. and Srivastava, P.S. 1995.*In-vitro* micro-propagation of *Moringa pterygosperma*. Phytomorphology. 45: 253-261.

Mohansundaram, M. 1985. Six new species of *Aculis Keifer* (Eriophyidae: Acarina) from South India. Mysore Journal of Agricultural Sciences. 15 (1): 22-26.

Mohapatra, R.N. 2006. Studies on bio-ecology and some aspects of management of cashew stem and root borer, *Plocaederus ferrugineus* L. M.Sc.(Ag.) Thesis Orissa University of Agriculture and Technology. Bhubaneswar, India.

Mohapatra, R.N. and Jena, B.C. 2007. Biology of cashew stem and root borer, *Plocaederus ferrugineus* L. on different hosts. Journal of Entomology Research. 31 (2): 149-154.

Moravec, C.M., Bradford, K.J. and Laca, E.A. 2008. Water relations of drumstick tree seed (*Moringa oleifera*): Imbibition, desiccation and sorption isotherms. Seed Science and Technology. 36 (2): 311-324.

Morton, J.F. 1991. The horse radish tree, *Moringa pterygosperma* (Moringaceae).A boon to arid lands ? Economic Botany. 45: 318-333.

Mridha, M.A.U. 2015.Prospects of moringa cultivation in Saudi Arabia. Journal of Applied Environmental and Biological Sciences. 5 (3): 39-46.

Mughal, M.H., Ali, G., Srivastava, P.S. and Iqbal, M. 1999.Improvement of drumstick (*Moringa pterygosperma* Gaertn) – a unique source of food and medicine through tissue culture. Hamdard Med. 42: 37-42.

Muhl, Q.E., du Toit, E.S. and Robbertise, P.J. 2011.Adaptability of *Moringa oleifera* Lam (Horse radish) tree seedlings to three temperature regimes. American Journal of Plant Sciences. 2: 776-780.

Munj, A.Y., Patil, P.D., Desai, S.D. and Naik, K.V. 2001.Seasonal incidence of major pests of drumstick, *Moringa oleifera* L. Pestology. 25 (2): 32-34.

Munj, A.Y. Patil, P.D. and Godase, S.K. 1998a. Biology of drumstick leaf eating caterpillar, *Noorda blitealis* Walker. Pestology. 22 (2): 18-21.

Munj, A.Y., Patil P.D., Rajput, J.C. and Godase, S.K. 1998b. Evaluation of some modern synthetic insecticides for the control of drumstick leaf eating caterpillar, *Noorda blitealis* Walk. Pestology. 22 (3): 24-26.

Murugavel, D.M.V. 2010.Drumstick, the magical medicinal tree. Intensive Agriculture, October-December 2010: 26-28.

Muthukrishnan, C.R. and Seemanthini, R. 1974. Perennial vegetables for your kitchen gardens.Indian Horticulture. 19 (2): 11-12.

Muthukrishnan, N. 2009.Integrated pest management studies. Annual Report. Horticultural College and Research Institute, Tamil Nadu Agricultural University, Periakulam, Tamil Nadu.

Muthuswamy, S. 1954. The culture of drumstick *Moringa pterygosperma* (Gaertn) (syn. *Moringa oleifera* Lam) in South India. South Indian Horticulture. 2: 18-21.

Nadkerin, A.K. 1976. Indian medicinal material Popular Prakasham: Bombay: 810-816.

Nair, M.R.G.K. 1995. Pests of drumstick – Insects and Mites of Crops in India. Indian Council of Agricultural Research, New Delhi. pp. 404.

Naiya, I. and Kabir, J. 2007. Influence of Virosil Agro on storage behaviour of drumstick (*Moringa oleifera* Lam) under high humidity conditions. Crop Research. 34: 143-145.

Narayana, C.K., Roy, S.K., and Pal, B.K. 1989. Use of high molecular weight high density polyethylene film (HM Film) in transit of mango (cv. Beneshan) III International symposium on mango hold at Darwin, Australia, November 1989. P. 89 (Abstract).

Narendra, M. 2007. Seed viability studies in drumstick (*Moringa oleifera* Lamk.) M.Sc. (Ag) Thesis. University of Agricultural Sciences, Dharwad, Karnataka, India. Karnataka Journal of Agricultural Sciences. 20 (4): 914 (Abstract).

Nautiyal, B.P., and Vanhataraman, K.G. 1987. Moringa (Drumstick) – an ideal tree for social forestry: Growing conditions and uses-part I. My Forest. 28(1): 53-56.

Ndubuaku, U.M. Ndubuaku, T.C.N. and Ndubuaku, N.E. 2014. Yield characteristics of *Moringa oleifera* across different ecologies in Nigeria as an index of its adaptation to climate change. Sustainable Agriculture Research. 3(1): 95-100.

Nigam, J.K. and Mishra, A.C. 2003. Drumstick: A Miracle tree. Indian Farmers' Digest. 36 (7-8): 29-30.

Nouman, W., Basra, S.M.A., Siddiqui, M.T., Yashmeen, A., Gull, T., Alcayde, M.A.C. 2014.Potential of *Moringa oleifera* L. as livestock fodder crop – a Review. Turkish Journal of Agriculture Forum. 38: 1-14.

Nouman, W., Siddiqi, M.T. and Basra, S.M.A. 2012a.*Moringa oleifera* leaf extract: an innovative priming tool for range land grasses. Turkish Journal of Agricultural Forum. 36: 65-75.

Nouman, W., Siddiqui, M.T., Basra, S.M.A. Afzal, I. and Urrehaman, H. 2012b.Enhancement of emergence potential and stand establishment of *Moringa oleifera* Lam. by seed priming. Turkish Journal of Agricultural Forum. 36: 227-235.

Odee, D. 1998. Forest biotechnology research in dry lands of Kenya: The development of *Moringa* species. Dry land Biodiversity. 2: 7-8.

Ojiako, F.O., Enwere, E.O., Dialoke, S.A., Ihejirika, G.A., Adikuru, N.C. and Okator, O.E. 2012. Nursery insect pests of *Moringa oleifera* Lam in Owerri Area Imo State, Nigeria. International Journal of Agriculture and Rural Development. 15 (3): 1322-1328.

Oliveira Jr. S.de., Souto, J.S., Santos, R.V., dos; Souto, P.C. and Major Junior, S.G.S. 2009.Fertilization with different manures in the cultivation of moringa (*Moringa oleifera* Lam.) Revista verde de Agroecologia e Disen

Volvimento Sustentarel. http://www.gvaa.com.br/revista/index.php/ RVAPS/art.

Olson, M.E. 2014. Spider mites love moringa. The international Moringa germplasm collection. National University, Mexico. (http://moringaceae. org/image-moringa-blog **spider mites-love-moringa.**

Padmavathi, A.S. and Manohara Prasad, D. 2015. Hints for moringa cultivation. Annadata, October 2005: 46-47.

Palada, M.C. 1996. Moringa (*Moringa oleifera* Lam.): A versatile tree crop with horticultural potential in the sub-tropical United States, Hort Science. 31(5): 795-797.

Palada, M.C. and Chang, L.C. 2003. Suggested cultural practices for moringa. International Cooperator's Guide (T.Kalbed.) AVRDC – Asian Vegetables Research and Development Centre, Shanhua, Taiwan, ROC, 03-545: 1-5.

Palada, M.C. Chang, L-C., Yang, R.Y. and Engle, L.M. 2007.Introduction and varietal screening of drumstick tree (*Moringa* spp) for Horticultural traits and adaptation in Taiwan. Acta Horticulturae. 752: 249-253.

Palanisamy, V., Balakrishnan, K., Karivaratharaju, T.V. and Arumugam R. 1995. Influence of seed treatments and containers on the viability of annual moringa seeds. South Indian Horticulture. 43 (1&2): 42-43.

Palanisamy, V., Kumaresan, K., Jayabharati, M. and Karivaratharaju, T.V. 1985. Studies on seed development and maturation in Annual moringa.Vegetable Science. 12 (2): 74-78.

Palgrave, K.C. 1983. Trees of Southern Africa.Struik, Cape Town, South Africa.

Paliwal, R., Sharma, V. and Pracheta, 2011a. A review on Horse radish tree (*Moringa oleifera*): A multipurpose tree with high economic and commercial importance. Asian Journal of Biotechnology. 3: 317-328.

Paliwal, R, Sharma, V., Pracheta and Sharma, S. 2011b. Elucidation of free radical scavenging and antioxidant activities of hydroethanolic extract of *Moringa oleifera* pods. Research Journal of Pharmalogical Technology. 4: 566-571.

Panda, A.K., Rama Rao, S.V. and Raju, M.V.L.N. 2008. Poultry manure – a potential bio-fertilizer for crops. Indian Farming. 58 (6): 13-14.

Pande, A. and Ghate, A. 1998.A new wilt disease of wild moringa (*Moringa concanensis*).Journal Tof Economic and Taxonomic Botany. 22 (2): 423-425.

Pande, V.C., Viswakarma, A.K., Singh, H.B. and Tiwari, S.P. 2013. Drumstick based Agri-Horticulture systems for profitability and resource conservation. Indian Farming. 63(9): 15-17.

Parr, J.F. and Colacicco, D. 1987. Energy in plant nutrition and pest control. In: Energy in World Agriculture. Elsevier Science Publishers. 2: 81-129.

Parrotta, J.A. 2001. Healing plants of peninsular India. CABI Publishing, Wallingford, UK and New York, USA: Pp. 56.

Parrotta, J.A. 2009. *Moringa oleifera* in Enzyklopadie der Holzgerwachse, Handbuch und Atlas der dendrology. A. Roloff, H. Weisgerber, U.Lang, and B. Stimm (eds) (Wein heim: Wiley – VCH verlog GmBH & Co KGAA), 40: 1-8.

Parthasarthy, P., Babu, S.R., Subramanian, S., Rabindra, R.J. and Rajasekharan, B. 2004. Comparative Toxicity of certain newer insecticides to the moringa hairy caterpillar, *Eupterote mollifera* (Walker) and its economics. Pest management in Horticultural Ecosysttems. 10 (2): 173-178.

Pasupathy, P. and Manvel, W.W. 2000.Moringa stem borer.Hindu. 09/11/2000. www.thehindu.com/2000/11/09/stories/08090421.htm

Patel, B.P., Radadia, G.G., and Pandya, H.V. 2010. Biology of leaf eating caterpillar, *Noorda blitealis* Walk on drumstick, *Moringa oleifera* L. Insect Environment. 16 (3): 135-138.

Peter, K.V. 1978. Drumstick – a rare Indian tree. Indian Farmer's Digest. 11:12.

Peter, K.V. 1979. Drumstick: A multi-purpose vegetable. Indian Horticulture. January – March 1979: 17-18.

Pillai, K.S., Saradamma and Nair, M.R.G.K., 1979. *Helopeltis antonii* Sign. As a pest of *Moringa oleifera*. Current Science. 49: 288-289.

Ponnappa, K.M. 1968. Some interesting fungi.III on typha Proceedings of Indian Academy of Science, India. Section, B. 68: 175-180.

Ponnuswami, V. 2012."Advances in production of Moringa (All India Coordinated Project on Vegetable crops). Tamil Nadu Agricultural University, Periakulam, Tamil Nadu.

Prabhakar, M. and Hebbar, S.S. 2007.Studies on organic production technology of annual drumstick in a semi-arid agro-ecosystem. Acta Horticulture. 752: 345-348.

Prabhu, M.J. 2009. A farmer's experimentation leads to a highly popular drumstick variety. The Hindu. January 29, 2009. Updated September 16, 2010.

Prasanth, P. and Srrnivasarao, A. 2004.Hints for moringa farming Rhthe Raju, Enadu 20 July 2004. 4.

Pugalendhi, L. and Swaminathan, V. 2010. Research on moringa in Tamil Nadu Agricultural University (In: Moringa. A compendium of brainstorming session on moringa held at Coimbatore organized by Dept.

of Vegetable Crops, TNAU, Coimbatore and AICRP on Vegetables Crops, IIVR, Varanasi, Ed. Pugalendhi, L., Singh, B., Natarajan, S and Makesh, S. 2010.

Punitha, P., Dabas, J.P.S., Sharma, N. and Joshi, P. 2019. Cultivating moringa: A miracle tree for health and well being .Indian Farmers' Digest, 52 (7): 33-38.

Radovich, T. 2009. Farm and Forestry Production, marketing profile for moringa (*Moringa oleifera*). In: Elevich, C.R. (ed) Specialty crops for Pacific Island Agro-forestry. Permanent Agriculture Resources (PAR) Hawaii (http://agroforestry.net.scps.)

Radovich, T.J.K. 2009. Arbuscular Mycorrhizal dependency of three Moringa Genotypes. Proceedings of the 2009 Annual conference of the American Society for Horticultural Science. (Millenium Hotel, St Louis) Ashs.Confex.Com/ashs/2009/... paper 1425.html.

Ragumoorthi, K.N. 1996. Bio-ecology, host plant resistance and management of moringa fruit fly *Gitona distigma* Meigen (Diptera: Drosophilidae), Ph.D. Thesis, Tamil Nadu Agricultural University, Coimbatore, TNAU, India.

Ragumoorthi, K.N. and Armugam, R. 1992. Control of moringa fruit fly *Gitona* sp. and leaf caterpillar, *Noorda blitealis* with insecticides and botanicals.Indian Journal of Plant Protection. 20 (1): 61-65.

Ragumoorthi, K.N., Selvaraj, K.N. and Subba Rao, P.V. 1998. Assessment of economic injure level (EIL) for moringa fruit fly, *Gitona distigma* (Meigon) (Diptera: Drosophilidae). In: Advances in IPM for Horticultural crops (eds) Parvatha Reddy et al., AAPMHE, Bangalore. pp. 137-139.

Ragumoorthi, K.N. and Subba Rao, P.V. 1997. First report of a palaeretic species on moringa fruit fly *Gitona distigma* (Meigen) in India, Pestology. 21 (9): 50-53.

Ragumoorthi, K.N. and Subba Rao, P.V. 1998. IPM for moringa fruit fly *Gitona distigma* (Meigen). Proceedings of International Symposium on Pest Management of Horticultural crops, Bangalore (Abstract).

Rahman, M.Z., Saud, Z.A. and Absar, N. 2001. A comparative investigation on the physio-chemical properties and some selected enzymes content in the flesh of healthy and *Rhizopus stolonifer* infected moringa fruit. Indian Phytopathology. 54 (3): 293-298.

Rai, N. and Yadav, D.S. 2005. Chapter 8.Perennial Vegetables. In: Advances in vegetable production. Research Book Centre, New Delhi. Pp. 645-649.

Raja, S. Appa Rao, V.V., Bagle, B.G. and More, T.A. 2008.Annual Report of Central Institute for Arid Horticulture, Bikaner, Rajasthan, India.

Raja, S., Bagle, B.G. and Dhander, D.G. 2006.Efficiency of different intercropping combinations with drumstick under dry land conditions.8th Agricultural Scientists and Farmers Congress. Banaras Hindu University, Varanasi. 21-22, February, 2006. p. 124.

Rajakrishnamurthy, V., Santhanabosu, S., Duraisamy, V.K. and Rajagopal, A. 1994. Drip irrigation in annual moringa. Madras Agricultural Journal. 81 (12): 678-679.

Rajamony, L., Mathew, S.K. and Celine, V.A. 2004. Flowering and fruiting patterns of drumstick in humid tropics. First Indian Horticulture Congress, 6-9 November 2004, New Delhi, Abstract. 7.95: page 58.

Rajangam, J., Azahaki Manavalan, R.S. Thangaraj, T., Vijayakumar, A. and Muthukrishnan, N. 2001.Status of production and utilization of moringa in Southern India. In: Development potential for moringa products. Workshop proceedings October 29 to November 2, 2001.Dares Salaam, Tanzania. In the "Miracle Tree / The multipurpose attributes of moringa" (Ed. L.J. Fuglie) CTA, USA.

Rajeswari, R and Kader, M.M. 2004. Effect of integrated nutrient management practices on growth, flowering and yield of annual drumstick cv. PKM-1. South Indian Horticulture. 52 (1-6): 194-202.

Ramachandran C; Peter, K.V. and Gopala Krishnan, P.K. 1980. Drumstick (*Moringa oleifera*): A multipurpose Indian Vegetable. Economic Botany. 34: 276-283.

Ramesh Kumar, A., Prabhu M., Ponnuswami, V., Lakshamanan, V and Nithyadevi, A. 2014. Scientific seed production techniques in moringa. Agricultural Reviews. 35 (1): 69-73.

Rangappa, T. 1980. Anaga Anaga Munaga Chettu, Annadata. May 1980: 20-22.

Ratnadas, S, Par Allainakari – Moussa, Ousmane Salha, Halrous Minet, Joel and Seyfowye, Amodou.Asmaou. 2011. *Noorda blitealis* Walker un ravageur dn moringa an Niger (Lepidoptera: Crambidae). Bulletin de la Societe entomologigue de France. 116 (4): 401-404.

Ravindran, C. 2013. Growing Moringa: Issues and constraints.www.teca.fao.org/discussion/growing moringa: issues and constraints.

Reddy, R.V.S.K., Reddy, M.T. and Babu, J.D. 2010.Status of moringa cultivation in Andhra Pradesh. In: L.Pugalendhi, Singh, B., Natarajan, S and Makesh, S (ea) Moringa. A compendium of brainstorming sessions on moringa 23-09-2010.TNAU, Coimbatore, TN, India. Pp. 110-112.

Reghapathy et al., 1991. Fruit fly threat to drumstick. The Hindu.dt. 13/02/1991.

Riad, S.R. El-Mohamedy; Abdalla, A.M. and Adam, S.M. 2014, Preliminary studies on response of *Moringa oleifera* plants to infection with some soil borne plant pathogenic fungi: International Journal of Current Microbiology and Applied Sciences. 3 (12): 389-397.

Riyathong, T., Dheeranupattana, S., Palaee, J. and Shank, L. 2010. Shoot multiplication and plant regeneration from *in-vitro* culture of drumstick tree (*Moringa oleifera* Lam.), In: Proceedings of the 8th International Symposium on Biocontrol and Biotechnology. Kung Mangkut's Institute of Technology Lad Krabang and Khan Kaen University, Nangkhai Campus, Thailand, 99-104.

Roe, N.E., Stofella, P.J. and Greatz, D.A. 1997. Composts from various municipal solid wastes, feed stock affect vegetable crops II. Growth, yields and fruit quality.Journal of the American Society for Horticultural Sciences. 122 (3): 433-437.

Sadashakti, A. 1995. Studies on combining ability, heterosis and prediction model for ideotype in annual moringa (*Moringa pterygosperma* Gaertn). Ph.D thesis, Tamil Nadu Agricultural University, Coimbatore, India.

Saha, T., Nithya, C. and Santhosh Kumar. 2014. Integrated Pest Management approaches for the insect pests of moringa (*Moringa oleifera*): Practices Rashtirya Krishi. 9 (2): 33-35.

Saini, R.K., Shetty, N.P., Giridhar, P. and Ravishankar, G.A. 2012. Rapid *in-vitro* regeneration method for *Moringa oleifera* and performance evaluation of field grown nutritionally enriched tissue cultured plants. Biotechnology. 2(3): 18-192. http://doi.org.10.1.1007/S13205-012-0045-9,

Saleh, S.A., Zaki, M.F., Tantway, A.S. and Salama, Y.A. 2016.Response of artichoke productivity to different proportions of nitrogen and potassium fertilizers. International Journal of Chemical Technology Research. 9(3): 25-33.

Sanchez, N.R., Ledin, S. and Ledin, I. 2006.Biomass production and chemical composition of *Moringa oleifera* under different management regimes in Nicargua. Agro-forestry Systems. 66: 231-242.   DOI10.1007/s10457-005-8847-4.

Sangeetha, G.K. and Ramani, N. 2007. Biological studies of *Tetranychus neocaladonicus* Andre(Acari: Tetranychidae) infesting *Moringa oleifera* Lam. Bulletin of Pure and Applied Sciences. 26 (2): 51-57.

Sangeetha V., Swaminathan, V., Beaulah, A., Rajkumar A. and Venkatesan, K. 2017. Effect of time of harvest, method of harvest and pre-packaging, calcium chloride treatments on shelf life and quality of moringa (*Moringa oleifera* Lam.) cv. PKM-1.International Journal of Current Microbiology and Applied Sciences. 6 (4): 212-221.

Sanjay Singh, 2010. Research on moringa at CIAH, Godhra, In: Pugalendhi, L., Singh, B., Natarajan, S. and Makesh, S. (ed). Moringa. A compendium of brainstorming session on moringa 23-09-2010, TNAU, Coimbatore, TN, India.

Saroja, P.L and Dwivedi, S.K. 1993. Moringa: A wonder tree. Indian Farmers' Digest. 26-27: 17-18.

Satti, A.A., Nasr, O.E.L., Fadelmula, A. and Ali, F.E. 2013. New record and preliminary bio-ecological studies of the leaf caterpillar, *Noorda bliteelis* Walker (Lepidoptera: Pyralidae) in Sudan. International Journal of Science and Nature. 4 (1): 57-62.

Satyagopal, K. et al., 2014. AESA based IPM for drumstick. National Institute of Health Management, Rajendranagar, Hyderabad – 500 030, India.

Seemanthini, R. 1964. A study of practices and problems in the cultivation of some perennial vegetables in Madras State. South Indian Horticulture. 12: 1-5.

Selvi, A. and Varadharaju, N. 2016. Controlled atmosphere storage of moringa (*Moringa oleifera*) pods, In: S. Narorro; D.S. Jayas. and K. Alagusundaram (eds) Proceedings of the 10th International Conference on Controlled Atmosphere and Fumigation in Stored Products (CAF 2016). CAF Permanent Committee Secretariat, Winniped, Canada, pp. 62-66.

Selvi, C. and Muthukrishnan, N. 2009. Biological characteristics of moringa bud worm, *Noorda moringae* Walk. on different annual moringa accessions. Indian Journal of Entomology. 71 (3): 269-271.

Shahzad, U., Jaskani, Md. J., Ahmad, S. and Awan, F.S. 2014. Optimization of the micro-cloning systems of threatened *Moringa oleifera* Lam. Pakistan Journal of Agricultural Sciences. 51 (2): 449-457.

Shamila, K., Harsh, N.S.K. and Joshi, K.C. 1996. Pathogenic fungus of *Noorda bliteelis* Walk.(Lepidoptera: Pyralidae) a major pest of *Moringa oleifera* Lamk. Indian Journal of Forestry. 19: 94-96.

Shamila, K. and Joshi, K.C. 1997. Efficacy of Foliar spraying of three varietal strains of *Bacillus thuringiensis* against the moringa defoliator *Noorda bliteelis* Tanss. (Lepidoptera: Pyralidae). Indian Journal of Plant Protection. 25 (1): 65-66.

Shanmugavelu, P., Paddappaiah, R.S. and Francis, K. 2003. Experimental testing of two agri-silvi-cultural models for semi-arid regions.Journal of Sustainable Forestry. 17 (4): 91-98.

Sharma, G.K. and Raina, V. 1982. Propagation Techniques of *Moringa oleifera* Lam. In: Improvement of Forest Biomass: Proceedings of a Symposium (Ed. P.K. Khosla) Solan, India. Pp. 175-181.

Sharma, N.D. and Joshi, A.C. 1977. Mycoflora associated with stored and freshly harvested rhizomas of *Zingibar officinale* Rose. Geobios. 4: 218-219.

Sharma, R. and Srivastava, R. 2008. Biofertilizer: From accessity to necessity. Agrobios Newsletter. VII (5): 50-52.

Shareef, Md., Kshirsagar, R.B., Sawate, A.R., Sayed, Z., Waghaye, S.Y. and Shailesh, V. 2019.Studies on effects of different chemical treatments on wholesomeness and post-harvest storage qualities of drumstick (*Moringa oleifera*) pods. International Journal of Chemical Studies. 7 (2): 329-333.

Sherkar, 1993.Republic of Germany. Pp. 334-337. Drumstick.The Bail Journal. 13 (2): 121.

Shokoohmand, A. and Drew, R.A. 2013. Micropropagation of *Moringa oleifera* Acta Horticulturae.988: 149-153.

Singh.H.P. 2011. Health management: exploring Research and Development of Moringa for nutrition and health care. Indian Horticulture. 56 (1): 3-8.

Singh, K.V., Datta, S., and Karmakar, P. 2014. Drumstick: the Wonder perennial vegetable for health management. Indian Farmers' Digest. 47(2): 36-37.

Sivagami, R. and David, B.V. 1968. Some insect pests of moringa (*Moringa oleifera* Lam) in South India. South Indian Horticulture. 16: 69-71.

Sivasubramanian, K. 1996. Studies on certain aspects of seed quality in moringa. Ph.D. thesis submitted to Tamil Nadu Agricultural University, Coimbatore, Tamil Nadu, India.

Sivasubramanian, K. and Thiagarajan, C.P. 1997. Storage potential of moringa seeds.Madras Agricultural Journal. 84 (10): 618-620.

Somali, M.A., Bajnedi, M.A., and Al-Faimani, S.S. 1984.Chemical composition and characteristics of *Moringa peregrina* seed and seed oil. Journal of American oil Chemists Society. 61: 85-86.

Sredevi, M., Mallaiah, K.V., and Yadav, N.C.S. 2007. Phosphate solubilization by *Rhizobium* isolates from *Crotalaria* species. Journal of Plant Science. 2: 635-639.

Sreenivasa Rao, G and Hari Prasad Rao, N. 1998. Better Management practices in moringa.Annadata. March 1998. Pp. 20.

Srinivas, P. 2017. Nation's current affairs: Telangana: Farmers switch to drumstick to tap demand from health freaks. Deccan Chronical.07.05.2017.

Staff, K.J. 2017. 2 lakhs in 6 months from drumsticks: Farmer's success www.krishijagaran.com>successstory>2 lakhs-in-6-months-from drumsticks-farmers-success.

Stanley, V.A. 2019. 3, 20,000 per acre... Profit even during drought from country moringa.www.scribd.com>document>3,20,000-per-acre-profiteven-during-drought-from-country-moringa.

Stephenson, K.K. and Fahey, J.W. 2004.Development of tissue culture methods for the rescue and propagation of endangered *Moringa* spp. germplasm. Economic Botany. 58(spl.1): S116-S124.

Stofella, P.V., Roe, M., Ozeres, Hampton and Greatz, D.A. 1997. Utilization of organic waste compost in vegetable production systems in Asia, (ed) R.A. Moris, Asian Vegetable Research and Development Centre, Shanhua, Taiwan.

Subba Rao, N.S. 2001. Soil Microbiology. Science Publishers, Inc. Enfield, New Hamphshire, USA, 407

Subramanian, T.R. 1965. A note on weevils damaging moringa. Indian Journal of Entomology. 27: 485-486.

Sundararaj, J.S. Muthuswamy, S., Shanmugavelu, K.G. and Balakrishanan, R. 1970., A guide on Horticulture 2nd edition, Velanpathippagam, Coimbatore, Pp. 261-262.

Sunthanapandian, I.R., Sambandamurthy, S. and Irulappan, I. 1989. Variation in seedlings population of annual moringa (*Moringa pterygosperma* Gaertn) South Indian Horticulture. 37: 300-302

Suresh Babu, K.V. and George, T.E. 2010. Research on moringa in Keral Agricultural University (In: Moringa A compendium of brain storming Session on moringa. L. Pugalendhi, B. Singh, S. Natarajan and S. Makesh (Eds) 2010. Organized by Dept. of Vegetables crops, TNAU, Coimbatore and AICRP on Vegetables crops, IIVR, Varanasi, India. 92-99.

Swati, B, Kamal; S, Kirad, A.K; Sharma and Mishra P.K. 2013. Standardization of propagation methods in drumstick cv. PKM-1.Nature and Science.11 (1): 141-143. www.sciencepub.net / nature / ns1101-022-15115ns1101.141-143pdf.

Tedonkeng, P.E., Boukila, B., Solefack, M.M.C, Kana, J.R., Tendonkeng, F. and Tonfack, L.B. 2004. Germination potential of *Moringa oleifera* Lam. under different treatments in Dschang in the high lands of Western Cameroon. Journal of Cameroon Academy of Science. 4: 199-203.

Teketay, D. and Demel, T. 1995. The effect of temperature on the germination of *Moringa stenopetala* a multipurpose tree.Tropical Ecology. 36 (1): 49-57.

Tenaye, A., Geta, E. and Hebana, E. 2009. A multipurpose cabbage tree (*Moringa stenopetala*) production, utilization and marketing in SNNPR. Ethiopia. Acta Horticulturae. 806 (Vol.1): 115-120.

Tewari, S.K., Pandey, A. and Kumar, V. 2003. Moringa: A useful agroforestry tree. Indian Farmers' digest, 37 (12): 13-14.

Thamburaj, S. and Natarajan, S. 2010. Propagation techniques in moringa (In: Moringa, A compendium of brainstorming session on moringa. Ed. L. Pugalendhi, B. Singh, S; Natarajan and S. Makesh 2010) organized by Dept. of Vegetable Crops, TNAU, Coimbatore and AICRP on Vegetable Crops, IIVR, Varanasi, India. Pp 113-118.

Ullasa, B.A. and Rawal, R.D. 1984. *Papaver rhoeas* and *Moringa oleifera*, two new hosts of papaya powdery mildew. Current Science, 53 (14): 754-755.

Umborter, Q. 2000. "World Dictionary of plant names, common names and scientific names". Eponyms and Etymology. 3: 1251-1731.

Usha Rani, B., Suresh, K. and Sundaram, R. 2010. Major insect pests of moringa and their management. In: Moringa a Natures Gift (Eds. Beaulah et al., 2010). Tamil Nadu Agricultural University, Coimbatore. Pp 54-59.

Vadivel, E. 2010. Systems of moringa cultivation in Tamil Nadu. Pp. 134-138. In: L. Pugalendhi, B. Singh, S. Natarajan and S. Makesh (eds). Moringa – A compendium of brain storming session on moringa 23rd September 2010. TNAU, Coimbatore, Dept. of Vegetable Crops, TNAU, Coimbatore, AICRP on Vegetables Crops. IIVR. Varanasi.

Valia, H.Z., Patil, V.K. and Patel, Z.N. 1993a. Effect of soil ESP on the growth and chemical composition of drumstick (*Moringa oleifera* Lamk).South Indian Horticulture. 41 (2): 84-90.

Valia, H.Z., Patil, V.K. Patel, Z.N., and Kapadia, P.K. 1993b. Physiological responses of drumstick (*Moringa oleifera* Lamk) to varying levels of ESP. Indian Journal of Plant Physiology. 36 (4): 261-262.

Veeraraghavthathan D., Jawaharlal, M. and Seemanthini, R. 1996.Drumstick Guide on vegetable cultivation. Suri Associates, Coimbatore, Tamil Nadu.

Verma, T.S. and Chaarasia, S.N.S. 1993-94. Drumstick (sainjana) for Better health and care. Indian Farmers' Digest. 26-27: 19-21.

Vijaykumar, K., Rubha, M.N., Manivasagan, M., Ramesh Babu, N.G. and Balaji, P. 2012. *Moringa oleifera*: The Nature's Gift. Universal Journal of Environmental Research and Technology, 2 (4): 203-209.

Vijaya Kumar, R.M. 2000.Studies on influence of months of sowing and growth regulators on annual moringa (*Moringa pterygosperma* Gaertn) Ph.D. Thesis, Tamil Nadu Agricultural University, Coimbatore, Tamil Nadu, India.

Vijayakumar, R.M. Srimathi, P., Vijayakumar, M. and Chezian, N. 1999. Influence of aging on the seed quality of annual moringa. South Indian Horticulture. 47 (1-6): 275-278.

Vijayakumar, R.M., Vijayakumar, M., Chezhian, N. Bangarsamy, V and Balasubramaniyan, T.N. 2002.Studies on the influence of months of sowing and growth regulating treatments on biometric observations in annual moringa cv. PKM-1.South Indian Horticulture. 50(4/6): 584-588.

Von Madell, H.J. 1986. Trees and shrubs of the Sahal, their characteristics and uses.GIZ, Eschbom, Germany. 525 pp.

Wang, H.F and Qiang, W. 2008. Establishment of regeneration system *in-vitro* for *Moringa oleifera* with stem (Chinese).Journal of Zhejiang Forestry Science and Technology. 28 (5): 40-43.

Wilson, S.B., Stofella, P.J. and Graetz, D.A. 2001. Use of compost as a media-amendment for containerized production of two subtropical perennials. Journal of Environmental Horticulture. 19 (1): 37-52.

Wu, L., Hallgren, S.W., Ferris, D.M. and Conway, K.E. 1999. Solid matrix priming to enhance germination of loblolly pine (*Pinus taeda*) seeds. Seed Science and Technology, 27: 251-261.

Xiang, S.Q., Liang, G, Guo, Q.G., Li, XL and He, O. 2007.Tissue culture and tetraploid induction of drumstick (Clinese) Journal of Tropical and Subtropical Botany. 15 (2): 141-146.

Youshef, Md. and Chouhan, S. 2009. Phytophagous mites of genus *Tetranychus dugour* (Acarina: Tetranychidae) infesting forest trees in India. My Forest. 45 (3): 273-281.

Zaki, M.F., Saleh, S.A., Tantawy, A.S. and El-Dewiny, C.Y. 2015.Effect of different rates of potassium fertilizer on the growth, productivity and quality of some broccoli cultivars under new reclaimed soil conditions. International Journal of Chemical Technology Research. 8 (12): 28-39.

Zaki, M.F., Tantawy, A.S., Saleh, S.A., and Helmy, Y.I. 2012.Effect of bio-fertilization and different levels of nitrogen sources on growth, yield components and head quality of two broccoli cultivars. Journal of Applied Sciences Research. 8(8): 3943-3960.

Index